Managing Biogas Plants
A Practical Guide

Green Chemistry and Chemical Engineering

Series Editor: Sunggyu Lee
Ohio University, Athens, Ohio, USA

Proton Exchange Membrane Fuel Cells: Contamination and Mitigation Strategies
Hui Li, Shanna Knights, Zheng Shi, John W. Van Zee, and Jiujun Zhang

Proton Exchange Membrane Fuel Cells: Materials Properties and Performance
David P. Wilkinson, Jiujun Zhang, Rob Hui, Jeffrey Fergus, and Xianguo Li

Solid Oxide Fuel Cells: Materials Properties and Performance
Jeffrey Fergus, Rob Hui, Xianguo Li, David P. Wilkinson, and Jiujun Zhang

Efficiency and Sustainability in the Energy and Chemical Industries: Scientific Principles and Case Studies, Second Edition
Krishnan Sankaranarayanan, Jakob de Swaan Arons, and Hedzer van der Kooi

Nuclear Hydrogen Production Handbook
Xing L. Yan and Ryutaro Hino

Magneto Luminous Chemical Vapor Deposition
Hirotsugu Yasuda

Carbon-Neutral Fuels and Energy Carriers
Nazim Z. Muradov and T. Nejat Veziroğlu

Oxide Semiconductors for Solar Energy Conversion: Titanium Dioxide
Janusz Nowotny

Lithium-Ion Batteries: Advanced Materials and Technologies
Xianxia Yuan, Hansan Liu, and Jiujun Zhang

Process Integration for Resource Conservation
Dominic C. Y. Foo

Chemicals from Biomass: Integrating Bioprocesses into Chemical Production Complexes for Sustainable Development
Debalina Sengupta and Ralph W. Pike

Hydrogen Safety
Fotis Rigas and Paul Amyotte

Biofuels and Bioenergy: Processes and Technologies
Sunggyu Lee and Y. T. Shah

Hydrogen Energy and Vehicle Systems
Scott E. Grasman

Integrated Biorefineries: Design, Analysis, and Optimization
Paul R. Stuart and Mahmoud M. El-Halwagi

Water for Energy and Fuel Production
Yatish T. Shah

Handbook of Alternative Fuel Technologies, Second Edition
Sunggyu Lee, James G. Speight, and Sudarshan K. Loyalka

Environmental Transport Phenomena
A. Eduardo Sáez and James C. Baygents

Resource Recovery to Approach Zero Municipal Waste
Mohammad J. Taherzadeh and Tobias Richards

Energy and Fuel Systems Integration
Yatish T. Shah

Sustainable Water Management and Technologies,
Two-Volume Set
Daniel H. Chen

Sustainable Water Management
Daniel H. Chen

Sustainable Water Technologies
Daniel H. Chen

The Water-Food-Energy Nexus: Processes,
Technologies, and Challenges
Iqbal M. Mujtaba, Rajagopalan Srinivasan, and Nimir Elbashir

Managing Biogas Plants: A Practical Guide
Mario Alejandro Rosato

Managing Biogas Plants
A Practical Guide

Mario Alejandro Rosato

CRC Press
Taylor & Francis Group
Boca Raton London New York

CRC Press is an imprint of the
Taylor & Francis Group, an **informa** business

CRC Press
Taylor & Francis Group
6000 Broken Sound Parkway NW, Suite 300
Boca Raton, FL 33487-2742

First issued in paperback 2020

© 2018 by Taylor & Francis Group, LLC
CRC Press is an imprint of Taylor & Francis Group, an Informa business

No claim to original U.S. Government works

ISBN-13: 978-1-138-62661-4 (hbk)
ISBN-13: 978-0-367-73587-6 (pbk)

Visit the Taylor & Francis Web site at
http://www.taylorandfrancis.com

and the CRC Press Web site at
http://www.crcpress.com

Contents

Foreword

The strong development of anaerobic digestion is rapidly changing the top positions, with countries with relatively less seniority in the industry such as the United States and Italy fighting for top ranking at global level, with outsiders such as China coming up strong.

This proves that the biogas industry is still far from having reached its full maturity, and the fact that what we don't know about it still outweighs what we have learnt so far.

Improving the efficiency of anaerobic digestion is now more important than ever. With increased attention to sustainability, the efficient use of soil, preserving its organic matter, as well as replacing chemical fertilizers with digestates, a deeper understanding of the biological processes behind biogas production is fundamental.

Mario Rosato is one of the pioneers in this sector, having spent many years in experimenting, disseminating, and training. More recently, he contributed with his experience to the workgroup drawing up the new Italian standard on the measurement of methane potential from humid anaerobic digestion.

This book is a summary of the topics covered in a number of training courses, and a guideline for conducting experiments, troubleshooting the more common digester issues, as well as to assessing the actual methane potential of given biomasses and, ultimately, their commercial value.

I first met Mario several years ago, still at the beginnings of the biogas boom, as he was fumbling with pipettes and brewing liquids, attended his highly appreciated lectures, where the clarity of his exposition testified the solidity of his practical experience.

This book will be helpful to anyone with a serious interest in learning more about anaerobic digestion, its key biological parameters and diagnostics, as well as in conducting independent experiments.

Piero Mattirolo
President of Distretto Agroenergetico Lombardo
(Lombard Agroenergy District)

Member of the Biogas/Biomethane Technical Committee of
FIPER—Federazione Italiana Produttori Energie Rinnovabili
(Italian Federation of Renewable Energy Producers)

Preface

For purely didactic reasons, a number of products and commercial brands will be quoted throughout this book. The inclusion of said references was a consequence of the simple fact that either the author had direct experience in the use of said products or that publishable material or information was available (including authorization to publish pictures, tables, etc.). Consequently, the said references do not constitute a recommendation for their purchase or an opinion as to their quality. All brands quoted are marked with ® and belong exclusively to their owners.

General Scope of This Book

There are three mutually independent but complementary strategies to optimize biogas plants:

1. *Optimum design of the plant*: The design of an existing biogas plant is not always the optimum. There are many reasons for this, but the most common are the following:
   a. The standardization of the components—with the aim of obtaining economies of scale as a function of the plant's size which the manufacturer considers to be optimal for his market conditions.
   b. Changes in norms and regulations after the plant's construction.
   c. Unforeseen variations in feedstock prices.
   d. Sometimes—more often than one could expect—the ignorance and superficiality of the designer.
2. *Increasing the net yield of methane for mass unit of feedstock.* Sometimes a biogas plant may have been specially designed to digest a given substrate. Even in such condition, it is always possible to extract more methane from the feedstock with the help of certain techniques, for instance, thermal or chemical pretreatment, milling or other ways of reducing particle size, sonication and addition of enzymes.
3. *Maximizing the methane yield per unit volume of the digester.* Another way to improve the performance of the biogas plant is to vary the parameters of the digestion process, for instance, its temperature, the intensity of stirring, dilution of the feedstock, etc.

With respect to the strategies outlined above, this book has three goals:

1. Providing the reader with a clear idea about what parameters are worth to measure.
2. Defining what instruments and techniques are the most suitable for such measurements.
3. Providing some guidance for the correct planning of the laboratory tests, based on the available instruments and human resources, in order to maximize the accuracy of the results.

The author kept the theory to a minimum, leaving much more room for practical study cases, showing how to optimize a biogas plant in a quick and simple manner. The method, developed by the author and described in this manual, requires a small investment in suitable laboratory equipment. Such investment is marginal in comparison to the total cost of the biogas plant itself or to the operational cost of laboratory tests performed by external providers.

In spite of the primarily practical approach of the book, the readers should thoroughly read the theoretical sections in each chapter, as they are essential for understanding and deconstructing many of the myths and legends that are common among biogas plant managers. Most of these arise from a partial or incomplete interpretation of the data available in the scientific literature, while sometimes they are the product of marketing spin from companies exaggerating the performance of some product. By focusing on the day-to-day, practical problems of biogas plant management, this book is a valuable instrument for those readers who do not necessarily have a professional background in biological or chemical laboratory techniques. The reader will learn, step by step, how to perform routine tests using just a minimal kit of instruments and how to transform the practical notions contained in this book into a successful professional activity.

Final Caveat

The author gained his experience by operating in the Italian biogas market, where most of the biogas plants run on agricultural feedstock. The Italian biogas market (the second in Europe by installed power) is strongly influenced by German companies and policies. Agricultural biogas plants in Germany and Italy aim to the production of energy, since this is strongly subsidized there. In such contexts, optimization means, "getting the maximum amount of energy per ton of feedstock." Plants operating in other countries or in the wastewater or urban garbage treatment sectors may not have the same goal.

For instance, the optimization of an urban garbage biogas plant, which gets its income from municipal tipping fees, may sometimes require sacrificing part of its energy productivity for the sake of a quicker digestion process or a safer sanitation of the waste.

Germany is the European leader in industrial biogas plants construction and the third market in the world after USA and China. Nevertheless, this does not mean that German technology and tradition will be suitable for effectively digesting the feedstock available in the Mediterranean or in other geographic regions. Many multinational companies promote their products in Europe by sponsoring research in German universities, because the prestige of the latter is a kind of endorsement of their product's quality. Hence, quite often the European literature on biogas is biased by marketing conditionings and in some cases, it is self-referenced too. The author's approach to adopting techniques, recipes, and information from the biogas literature is just applying the Cartesian scientific method: test in your laboratory if other people's findings apply to your own case and check how accurate your tests are. Never test new methods, or feedstock, or additives directly in the biogas plant!

Each anaerobic digester is a unique ecosystem of microorganisms: bacteria, protozoans, archaea, yeasts, etc. The equilibrium between these is variable in time and hard, if not impossible, to predict. It cannot be taken for granted that a result published by a research group, or in the many examples contained in this book, will work or bring a benefit to the reader's digester in a given moment of time. The following list summarizes the author's advices:

1. Do not forget that every measure, even those taken by famous university laboratories, is subject to errors. The concept of "measure error" is still a taboo for most researchers and even the German norm VDI 4630/2014 contains a conceptual confusion between accuracy and precision. Hence, always be cautious about adopting biogas plant management methods based on tables.

2. Plan and conduct tests in order to know exactly how the biological process is running in your own anaerobic digester, in this precise moment. Tests you may have conducted in the past not necessary reflect the current state of the biological ecosystem or the quality of the current feedstock lot.

3. Measure the yield of your locally available feedstock, lot by lot, using inoculum taken from your own biogas plant. If you buy feedstock for your biogas plant, knowing its exact methane yield will allow you to better negotiate its price.

Acknowledgments

To my wife Giovanna, without whose patience and help this book would never have come to press.

To my students and customers, whose questions and comments have contributed to making this text clearer and more complete.

To Professor Jing Liu and all the staff of Bioprocess Control AB, for the photos and advice provided.

To Professor Raman Saravanane for the valuable advice and bibliographic material provided.

Latisana (Italy),
January 2017

About the Author

Mario A. Rosato is an electric–electronic and environmental engineer, as well as a scientific journalist. He built his first home-sized digester in Argentina when he was aged 16. The son of a university professor and researcher, since his childhood he learnt to employ sophisticated instruments in his favorite playground—his father's lab. At the age of 25, he obtained a scholarship to specialize in renewable energies in Italy. At 28, he chose to leave the academic research world and devoted himself to the development of industrial solutions. In 1990, he settled in Italy, and in 2000, he moved to Spain. In 2004, he became partner and scientific director of Bioenergia Aragonesa SL. In 2006, he patented the AFADS system, a multitrophic bioreactor for wastewater treatment. In 2009, he founded Sustainable Technologies SL in Barcelona. In 2010, he expanded his company's activities to Italy, where he became a professor in several private professional institutes. Since then he has trained more than hundred biogas plant managers in Italy and Spain. Mario Rosato installed a laboratory for applied research and routine anaerobic digestion tests in the Technologic Park of Pordenone in 2011. The same year he won The Economist's award for the best entrepreneurial idea for tackling global climate change, based on the production of biohydrogen and the cultivation of bamboo in a circular economy cycle. Few months later, he received the Green Vision Award from the Modus Vivendi magazine in Rome. His industrial solutions, based on original applied research, place him fourth in the worldwide Top Solvers list of Innocentive. com. In 2012, during the research project H2 Ocean, funded by the European Commission's seventh Framework Program, he developed a novel type of digester for marine biomass, specially conceived for offshore operations. In 2013, he won two international awards for the conceptual design of a domestic garbage digester, meant for low-income Indian families. Since 2014, he has been in charge of the column on bioenergy for agronotizie.it, a specialized e-zine for agronomists and farming professionals. In 2015, he became the member of the technical commission in charge of redacting the Italian norm on the biochemical methane potential (BMP) test protocol, his contribution being focused on error propagation analysis and improving both accuracy and precision.

1

Relevant Aspects for Optimizing the AD Process

1.1 What Is Anaerobic Digestion?

1.1.1 Theoretical Notions

Anaerobic digestion (AD) is the biological process leading to the decay of organic matter in the absence of air. Because air consists of 21% oxygen, the anaerobic process is sometimes called anoxic.

Several types of anaerobic fermentative processes exist in nature:

- Alcoholic fermentation, e.g., during the production of wine or beer.
- Lactic fermentation, e.g., during the production of *sauerkraut* (*choucroute*), corn silage, and soy sauce.
- Dark fermentation. It is a particular type of anaerobic fermentation that produces biohydrogen, and requires energy to be supplied as heat.
- Photofermentation. Certain bacteria and cyanobacteria produce biohydrogen, but only if energy is supplied to them as light.
- Butyric fermentation. It is a particular type of dark fermentation, leading to simultaneous production of hydrogen, butanol, and acetone.
- Propionic fermentation. It is the production of propionic acid by means of *Propionibacterium acidipropionici*, grown under strict anaerobic conditions using lactose as the carbon source, at pH 6.5 and temperature 30°C.
- Methanogenic fermentation. This process is discussed in detail in this book, although some of the techniques presented here can be employed to optimize different fermentative processes.

The organisms directly involved in producing methane are the *Archaea*, very similar to the *Bacteria*, but belong to another kingdom, and are probably the bacteria's ancestors.

The kingdom *Archaea* encompasses two subkingdoms: the *Crenarchaeota* and the *Euryarchaeota* (Figure 1.1).

The *Crenarchaeota* are called extremophile organisms, because they usually live in environments like submarine volcanoes, sulfur lakes, saline lakes, or even in deep rocks, where the temperature, or the pH, or the pressure, or the salinity reaches values that would be lethal for any other life forms. The use of these organisms in industrial applications began in 2014, as researchers found some promising species capable of producing ethanol directly from cellulose or repairing their own DNA when damaged by very intense radiations.

The *Euryarchaeota* in turn include all the species of methanogenic *Archaea* known until now. The biodiversity is very high within this subkingdom, because while some species thrive in the cold seabed of the Arctic, others prefer temperatures near 100°C; yet other species take their nourishment from hydrogen, which is toxic for other species that feed on acetic acid.

The AD process is a sequential combination of fermentative processes in which the last stage is the methanogenic fermentation. As of 2017, our knowledge of the AD processes can be sketched as shown in Figure 1.2.

The organic matter is composed of carbohydrates (compounds of carbon, hydrogen, and oxygen, namely, simple sugars and polysaccharides), proteins and amino acids (compounds of carbon, hydrogen, oxygen, and nitrogen), and lipids (compounds of carbon and hydrogen, with small proportions of oxygen). Furthermore, all organic substances contain a mineral fraction, called ash. The biological decomposition of carbohydrates, proteins, and lipids cannot be taken for granted: some substances like lignin (a very complex carbohydrate) are absolutely indigestible by anaerobic organisms. Other substances like keratin (the protein that composes hair and feathers) are

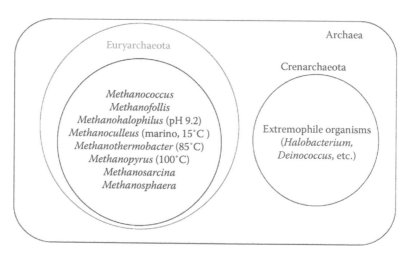

FIGURE 1.1
Situation of the methanogenic organisms within the kingdom *Archaea*.

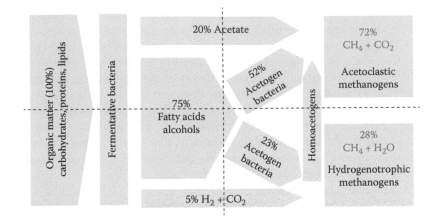

FIGURE 1.2
The phases of the AD process, according to Varnero Moreno (2011). (Graphic re-elaboration by the author.)

very difficult to degrade anaerobically. Generally, the more complex the substance, the higher will be the number of microorganisms involved in its degradation, and hence the longer the time for its complete AD. It is known that some pesticides, not degradable through aerobic processes and very difficult to degrade chemically, can be decomposed by anaerobic bacteria, although with very long times.

Observing Figure 1.2, we notice that the fermentative bacteria (true *Bacteria*, not *Archaea*) and other microorganisms (yeasts) are responsible for the first demolition step of the organic molecules. Assuming the total amount of organic matter brought for fermentation as 100%, 20% of it becomes acetic acid and reacts with the salts present in the fermentation medium, forming acetate that in turn feeds a group of methanogens, called acetoclastic methanogens. These latter are archaea and thrive by converting the acetate into methane (CH_4) and carbon dioxide (CO_2). Together with the acetic acid, the first fermentative stage transforms 5% of the initial organic matter into free hydrogen (H_2) and CO_2. Both gases, dissolved in the aqueous phase of the fermentation broth, feed another group of archaea, called hydrogenotrophic methanogens, that synthetize methane and water. The remaining 75% of the organic matter, partially degraded during the first fermentative stage, is composed of molecules of intermediate dimensions: volatile fatty acids (VFA), intermediate-chain fatty acids, alcohols, ketones, etc. The said intermediate molecules are in turn demolished by the acetogenic bacteria that produce more acetic acid and also hydrogen and carbon dioxide, thus feeding both the archaea groups and a third group of bacteria, called homoacetogens, which synthetize acetic acid from the hydrogen and carbon dioxide molecules. In general, when digesting "rich" organic substrates, 72% of the total amount of methane is because of the acetogenic and homoacetogenic activities of the bacteria, and both are followed by acetoclastic transformation. The remaining 28% results from the hydrogenotrophic path. Depending on the literature sources, the proportions quoted may vary slightly from author to author, with some of them estimating 70% of the total amount of methane produced through the first pathway and 30% through the second.

The popular wisdom says that "Man does not live by bread alone," and this applies to bacteria too. In the same way that superior animals need a balanced diet, also the bacteria in a biogas plant require certain attention to the preparation of their "menu." Fortunately, the ideal diet for a species or bacterial group is a bit easier to formulate than that for a cow or a person. If we consider the digester as an ecosystem of microorganisms, the ideal diet should just have a ratio between carbon and nitrogen (C/N) as close as possible to 30. Since phosphorus is also an essential element for life, the microorganisms' diet must take it into account, resulting in an optimum growth when the ratio of C/N/P is equal to 150/5/1. The reason why it is practically impossible to formulate the diet of the biogas plant based only on the C/N/P proportions is simply that the said elements must not only be present in the feedstock, but must also be in a form that is easy to assimilate by the microorganisms. Like people or animals, microorganisms are also healthy when they feed on light substances and, conversely, they can even die of indigestion if their diet is too "heavy." The *emoticons* in Table 1.1 help us to remember this important concept.

From the stoichiometric point of view, the reactions leading to the anaerobic degradation of pure substances are provided in Table 1.2.

The value δG_0 represents the total energy consumed (if negative) or produced (if positive) by the reaction. We can notice from Table 1.2 that all chemical reactions of the anaerobic degradation known to the present show a negative sign. This means that the reaction needs energy in the form of heat to take place, and hence it is said to be an endothermic reaction. Depending on the temperature of the heat supply, the AD process can be classified as:

- Psychrophilic, for T between 10°C and 20°C ± 2°C
- Mesophilic, for T between 20°C and 40°C ± 1°C
- Thermophilic, for T between 40°C and 60°C ± 0.5°C
- Extremophilic (over 60°C, when other kinds of *Archaea* and *Bacteria* produce mostly H_2, methanogens are inhibited)

TABLE 1.1

Digestibility of the Organic Substrates, from Best to Worst

Light carbohydrates (sugars, starch)	☺
Heavy carbohydrates (polysaccharides)	☺
Proteins	☺
Fats, soaps	☹

TABLE 1.2

Stoichiometry of the AD

No.	Reaction	δG_o [kJ/mol CH_4]
1	$4CH_3OH \rightarrow 3CH_4 + CO_2 + 2H_2O$	−106
2	$CH_3OH + H_2 \rightarrow CH_4 + H_2O$	−112.5
3	$4CH_3NH_2 + 2H_2O \rightarrow 3\ CH_4 + CO_2 + 4NH_3$	−76.7
4	$2(CH_3)_2NH + 2H_2O \rightarrow 3CH_4 + CO_2 + 2NH_3$	−74.8
5	$4(CH_3)_3N + 6H_2O \rightarrow 9CH_4 + 3CO_2 + 4NH_3$	−75.8
6	$2(CH_3)_2S + 2H_2O \rightarrow 3CH_4 + CO_2 + 2H_2S$	−52.1
7	$4(CH_3)SH + 2H_2O \rightarrow 3CH_4 + CO_2 + 4H_2S$	−51
8	$(CH_3)SH + H_2 \rightarrow CH_4 + H_2S$	−69.3
9	$4H_2 + CO_2 \rightarrow CH_4 + 2H_2O$	−130.4
10	$CH_3COO^- + H^+ \rightarrow CH_4 + CO_2$	−36
11	$4CO + 2H_2O \rightarrow CH_4 + 3CO_2$	−211

N.B.: Some authors define the range 40°C–80°C as thermophylic and do not define an extremophilic range. This is just a question of conventions and age of the publication, since extremophilic processes are a relatively new research field.

According to a study by Gerardi (2003), the optimum temperature ranges are different for each genus of *Archaea*, as summarized in Table 1.3.

Hence, according to the said research, the maximum biodiversity of *Archaea* species corresponds to a very narrow temperature range between

TABLE 1.3

Ideal Temperature Ranges for Each Genus of *Archaea*

Genus	20°C	25°C	30°C	35°C	40°C	45°C	50°C	55°C
Methanobacterium				**	*****	*****		
Methanobrevibacter				**	*****			
Methanosphaera				*****	*****			
Methanolobus				*****	*****			
Methanococcus				*****	*****			
Methanosarcina			*****	*****	*****			
Methanocorpusculum			*****	*****	*****			
Methanoculleus				*****	*****			
Methanogenium	*****	*****	*****	*****	*****			
Methanoplanus			*****	*****	*****			
Methanospirillum				*****	*****			
Methanococcoides			*****	*****				
Methanolobus				*****	*****			
Methanohalophilus				*****	*****	*****		
Methanohalobium							*****	*****
Methanosarcina							*****	*****

37°C and 40°C. The maximum biodiversity of the methanogenic ecosystem leads to a stable operation of the AD plant (biogas plant), less prone to "jumps" of the biogas production and more resilient to disturbing events (e.g., sudden changes in the composition of the digester's diet, excursions of the pH, and accidental input of inhibiting substances). The former concepts seem to justify the choice of many biogas plant constructors, who promote mesophilic processes as the most reliable in the market.

According to another study conducted by Dutch researchers, the maximum anaerobic degradation speed of the acetate corresponds to an incubation temperature of about 60°C–65°C, and the methanogenic activity varies with the fermentation temperature as shown in Figure 1.3.

The results of both studies seem to confirm the choice of some biogas plant constructors, who state that their thermophilic plants, operating at about 50°C, are the most efficient (more compact, because of their shorter hydraulic and solid retention times). We will analyze in depth both these (mesophilic vs. thermophilic) in the forthcoming chapters. At this stage, we can just anticipate that both are partially correct, because in practice nobody can demonstrate that one is in all cases better than the other. Remember that the AD is a sequential process and hence the overall digestion speed is limited by the slowest stage. The fermentative bacteria are in general mesophilic, whereas the *Archaea* tend to be thermophilic, although not all of them are. Hence, a digester that optimizes the living conditions of one microbial group will hamper the other and vice versa.

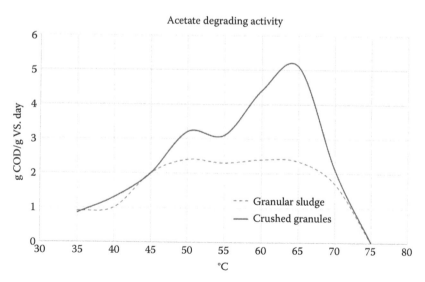

FIGURE 1.3
Anaerobic degradation speed of acetate as a function of temperature, measured on two different types of inoculums. (From Van Lier, J., et al., *Water Research*, 30(1), 199–207, 1996; graphics by the author.)

1.1.2 Practical Implications

From the former explanations, we can draw the following fundamental practical conclusions.

1.1.2.1 Complexity of the System

The AD is a very complex process, in which more than 1000 species of microorganisms must coexist, each of them having different vital needs. For instance, the fermentative *Bacteria* prefer acid (low) pH, while the *Archaea* grow well if the pH is kept as close as possible to neutrality. In other words, to manage a biogas plant means to become a "bacteria farmer." Such a job has some similarities to that of the pig or the chicken farmer. It is necessary to feed the bacteria in the digester with the correct diet and maintain the hygienic conditions of their ecosystem. Never forget that microorganisms are as sensitive to both excess and deficit of feedstock, to the presence of inhibiting pollutants, and to extreme temperature, as superior animals. The biggest difference between farming bacteria and farming animals is that the "fodder" for the digester can generally be a single substance that will be digested by an ecosystem composed of over thousand different species, some of which may be more or less efficient in degrading the said feedstock. A biogas plant is furthermore an ecosystem showing higher complexity and biodiversity than a natural park. Hence, the correct management of a biogas plant is not just "throwing waste into the digester." In contrast, it is necessary to calculate in each circumstance, as exactly as possible, which substrates and in which proportion must they be fed into the digester, so as to obtain the desired energy production. Furthermore, it is necessary to prepare each substrate in the best way before loading it into the digester, in order to enhance its total conversion into biogas, or at least its highest rate of conversion, since most substrates are not fully degradable (Figure 1.4).

1.1.2.2 Sequential Process

The AD is a process that follows a precise sequence of biochemical reactions. It is common sense that "a chain cannot be stronger than the weakest of its rings," so a single perturbation (e.g., feedstock containing antibiotics, sudden variations of temperature or pH, etc.), introduced at any trophic level, may lead the anaerobic process to collapse.

1.1.2.3 Multiparametric Process

For the same reasons already explained, there is *no simple and direct* way for monitoring the AD process, neither is it possible to conduct a census of the microorganisms to check which are active or not. Hence, it is necessary to

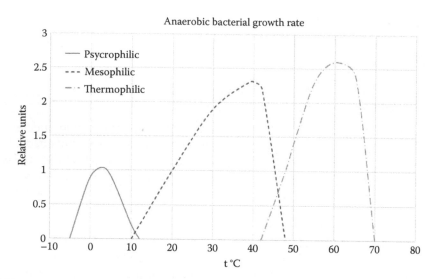

FIGURE 1.4

Relative reproductive activity of the different species of *Archaea*, according to Madigan and Martinko (2006), republished by Schnürer and Åsa (2010). (Graphics by the author.)

perform *indirect tests* to understand the population consistency and the biological activity of each group of microorganisms composing the "ecosystem" in the digester.

1.1.2.4 Two Degradation Paths of the Organic Matter

In many books and articles available from different sources, the AD process is still represented as exclusively limited to the acetogenic–homoacetogenic–acetoclastic path, ignoring the hydrogenotrophic path, which anyway contributes to 28% of all the theoretical methanogenic capability. This conception, at least 20 years old, still remains in the design of many biogas plants (single tank or two identical tanks in parallel), making them suboptimal from their construction. Since it is almost impossible to modify a biogas plant after it is already built, it becomes critical to determine, by means of small-scale tests, the best operational conditions for each reactor, settling for a compromise in the choice of the process parameters (T, pH, stirring, etc.), as will be discussed in the forthcoming chapters.

1.1.2.5 Influence of Temperature

Temperature is a factor of crucial importance, probably the one that has raised the largest number of "rural legends" among designers, builders, and managers of biogas plants. In practice, the ideal operation temperature of the biogas plant will depend on the composition of the bacterial ecosystem, considered as a whole. Consequently, it is too restrictive to decide the operation temperature only on the basis of the acetate degradation test, like the one shown in Figure 1.3, because the said test does not consider the 28% of the total methane production that comes from the hydrogenotrophic pathway. Nor is the biodiversity of the sole *Archaea* the best criterion to decide the working temperature of a digester—several other microorganisms must degrade the biomass before the *Archaea* are able to produce methane, and the said microorganisms may not necessarily require the same ideal temperature. Once again, each biogas plant is a singular case requiring some practical tests to find out the ideal operation temperature for that specific bacterial consortium.

1.1.2.6 Need of a Balanced Diet for the Microorganisms

The ideal C/N ratio is 30; nevertheless, not all the carbon or the nitrogen present in the feedstock is necessarily digestible. Consequently, it is possible to obtain good results by the AD of substrates having C/N ratios ranging from 10 to 90. Furthermore, potassium, calcium, magnesium, and other elements help, within some limits, to keep a healthy bacterial ecosystem, but there is much marketing hype on their importance, because some companies and biogas plant builders have built a business on selling mineral additives or "boosters." When the feedstock for a biogas plant consists of a mixture of different biomasses, it is very advisable to test different proportions so as to find out the optimum combination. It is not advisable to rely only on the elementary chemical composition of each biomass, as measured by animal feed laboratories or by Near InfraRed Spectrometry (NIRS) apparatus, nor on data from the scientific literature, because the anaerobic degradability of a given biomass depends also on other factors, like the feedstock's age and conservation method, eventual pretreatments, etc. A general idea on the C/N ratio of the different biomasses composing a mixture is anyway useful, at least to calculate the proportions that are nearer to the optimum, so as to minimize the amount of biological tests to be performed before deciding the digester's diet. Table 1.4 helps the reader to predefine the tests with logical criteria, and furthermore Section 1.2.2.6 provides a practical example of its use.

1.1.2.7 Criterion for the Selection of Feedstock for AD

The theoretical concepts summarized in Table 1.1 allow us to define a selection guide based on the quality of the most common agricultural biomasses, ordered from most to least digestible, as shown in Table 1.5.

1.2 Technical Nomenclature

Each branch of science or technique has its own jargon. Sometimes, the use of technical terms in a book, or the assumption that some technical conventions

TABLE 1.4

Approximate Contents of C, N, and C/N Ratios of Some Common Biomasses (% w.w. = Percent of Wet Weight)

Biomass	N (% w.w.)	C (% w.w.)	C/N
Corn (whole plant)	1.2	68	56.6
Cabbage	3.6	45	12.5
Grass (hay)	4	48	12
Alfalfa	2.5	40	16
Hay (leguminous)	1.6	40	25
Cow dung	1.6–1.8	30–40	17–25
Sheep dung	3.8	49–76	13–20
Stable dung (with straw)	0.8	22	27.4
Horse dung	2.3	57.5	25
Swine dung	2.8–3.8	23–38	6.2–13.7
Swine manure	0.4	4	10
Straw (oat)	0.5	40	80
Straw (wheat)	0.5	50	100
Straw (barley)	1	48	48
Hen dung (layers)	3.7–6.3	31–35	5–9.6
Tomatoes	3.3	41	12.5
Kitchen waste	1.9	54	28.6
Corn stalks	1.4	44	31
Clover	3	39	13

TABLE 1.5

Anaerobic Digestibility of the Most Common Biomasses

Corn, sugar beet, bread, molasses, sugars	Easily degradable, but unsustainable
Horse dung	Digestible
Cow dung	Digestible
Goat/sheep/rabbit dung	Digestible
Organic municipal solid waste (shredded)	Digestible
Agro-food waste (vegetables, fruits, extraction cake)	Digestible
Water plants (water hyacinth, duckweed)	Digestible, but not widely available
Fresh grass	Digestible
Hen dung (layers)	Low C/N ratio
Swine manure	Low C/N ratio
Slaughterhouse waste	Low C/N ratio
Corn cobs and stalks, wheat and oat straw	Digestible, high C/N ratio
Rice straw, canes, vegetal fiber, chicken dung (broilers)	Difficult digestion
Fats and soaps, hair, feathers	Slow digestion
Wood, sawdust, hooves, horns, etc.	Indigestible

are known to everybody, may lead to misunderstandings and disastrous results when attempting to apply the theory to practical cases.

Many words will be recurrent within this book, and so it is worth dedicating a few lines to define them so as to avoid misunderstandings.

1.2.1 Definitions

The following lines give exhaustive definitions of the various concepts and parameters that are useful for the correct management of the biogas plant. The presentation order follows the logic of the forthcoming chapters. Chapter 7 contains a summary of the definitions, in alphabetical order, presented as a glossary, for an easier consultation.

Substrate
Biomass, animal or vegetal, that will be employed as feedstock for AD.

Inoculum
Biomass containing living microorganisms, necessary to start and maintain the AD process.

Hydraulic Retention Time (HRT) or Solids Retention Time (SRT)
It is the average amount of time that a particle or small portion of matter remains inside the digester. By definition, HRT = Volume of digester (m^3)/ feedstock's flow (m^3/day).

Cellular Retention Time (CRT)
It is the average amount of time that bacteria, or a colony of microorganisms, remain inside the digester. The CRT not necessarily coincides with the SRT and depends on the type or family of digester.

W.W. = wet weight
It is the weight of a sample as it is when starting a test, i.e., including its moisture content.

Total solids (TS); dry weight (DW)
By definition, TS = WW − Moisture content.

VS = volatile solids
Fraction of the TS that is assumed as digestible. By definition, VS = TS − Ash content.

Ash
Mineral fraction contained by the organic matter.

Chemical Oxygen Demand (COD)
An indirect measure of the organic carbon quantity present in the biomass. The COD is a highly standardized, relatively quick chemical test.

Biological Oxygen Demand (BOD)
A measure of the organic carbon quantity present in the biomass that can be aerobically degraded. This parameter is seldom employed in the biogas

industry, because the aerobically degradable organic matter is not necessarily anaerobically degradable and vice versa.

Organic Load (OL) or Total Organic Load
The total daily mass of VS or COD introduced in the digester (kg or ton of VS or COD/day)

Organic Loading Rate (OLR)
The OL divided by the digester's volume (kg or ton of VS or COD/m^3 digester day)

Biochemical Methane Potential (BMP)
It is the net amount of methane produced by the fermentation of a mass unit of substrate in a given time. In general, the acronym BMP is followed by a suffix, indicating the duration of the test with which the BMP value was calculated. For instance, BMP_{20} means the total amount of methane produced during a test that lasted 20 days, per unit of mass. If BMP is not followed by a suffix, it is conventionally assumed that the test lasted 30 days. The correct unit for expressing the BMP is Nml/g of VS or Nl/kg of VS, or Nm^3/ton VS. N.B.: COD can also be employed instead of VS, especially when dealing with liquid substrates.

Nl or Nm^3
Volume occupied by a gas, in liters or m^3, under "normal conditions." By "normal" it indicates 0°C and atmospheric pressure equal to 101.3 kPa.

Sl or Sm^3
Standard cubic meters. This way of defining volumes is widely diffused in the natural gas industry but seldom employed in the biogas industry. In this context, "standard" means m^3 of gas at 15°C and 101.3 kPa. The conversion factor between both reference conditions is: 1 Nm^3 = 0.947 Sm^3.

Alkalinity (aka buffer capacity)
It is a measure of the resistance to the variations of pH of a given solution. The alkalinity is proportional to the concentration of carbonates and bicarbonates. There are several ways of expressing the buffer capacity; usually it is measured as equivalent mg of CO_3Ca/l of solution.

pH = hydrogen potential
It is a measure of the activity of the OH radical in aqueous medium, in a scale from 1 to 14. If pH = 7, the solution is neutral (distilled water), if pH < 7 the solution is acid (e.g., vinegar, beer, Coca Cola, lemon juice, hydrogen peroxide), and if pH > 7, the solution is said to be basic or alkaline (e.g., solution of caustic soda in water, tap water, sea water, ammonia, bleach, dishwasher soap).

Volatile Fatty Acids (VFA)/Total Alkalinity ratio (TA), aka FOS/TAC in the German and some European literature
Ratio between the total concentration of VFA (FOS in German) and the total alkalinity (total inorganic carbonates, TAC in German).

Lower Heating Value (LHV), aka Lower Calorific Value (LCV)
The LHV of pure methane is 9.94 kWh/Nm^3, whereas for biogas it is conventionally assumed to be 60% of the said value.

Sludge, slurry, and digestate
From a scientific point of view, slurry and sludge are the same thing—the organic matter under fermentation contained in the reactor during the AD process.

Digestate is the slurry resulting from the AD process. In a single-reactor biogas plant, the sludge is at the same time the digestate, while in a multiple-reactor biogas plant, the digestate is the sludge exiting from the last digester or from the storage tank.

Volatile Fatty Acids
Any organic acids (carboxylic acids) having less than six carbon atoms, i.e., formic acid (CH_2O_2), acetic acid ($C_2H_4O_2$), propionic acid ($C_3H_6O_2$), butyric acid ($C_4H_8O_2$), and valeric acid ($C_5H_{10}O_2$). The most relevant ones for the AD process are acetic acid (the preferred feedstock for acetoclastic *Archaea*), propionic acid, and butyric acid. The presence of formic and valeric acids beyond certain concentrations is an indicator of unbalances of the microbial ecosystem or of eventual problems in the AD process, because the said acids should not form under normal conditions of the organic matter degradation.

Oxidation Reduction Potential (ORP) aka redox potential
The ORP is defined as "the tendency of a chemical species (molecule or categories of molecules or radicals) to gain electrons and hence to suffer a reduction." The ORP's measure unit is millivolt (mV). It is a chemical magnitude very simple and economical to measure by means of electrodes, very similar to those employed for pH measures. The ORP is largely employed as the quality parameter of water in the environmental protection and wastewater treatment industry. Values between 0 and +2000 mV indicate aerobic activity. For example, values between 0 and 150 indicate a high bacterial activity consuming O_2, and hence contaminated water; values between 400 and 450 mV indicate good quality of water for a swimming pool; values from +600 mV indicate disinfection and potable water; sterilization of water with ozone brings the ORP values to over +800 mV. Values between 0 and –2000 mV indicate anaerobic processes. In the methanogenic AD process, the production of carbon dioxide is the result of an *oxidation* process of the organic carbon, while the production of CH_4 is a *reduction* process of the organic carbon. Hence, we can deduce that, during anaerobic fermentation, the reactions are extremely complex to monitor, since they encompass both oxidation and reduction of the organic carbon present in the digester. The literature indicates typical ORP values between –220 and –290 mV as those optimum to maintain a stable AD process (Blanc and Molof (1973, quoted by Lee Sung J.)).

1.2.2 Practical Applications and Numerical Examples

It is useful to dedicate some time to get familiar with the definitions given in Section 1.2.1, because quite often they will be referred to in this book. The reader is encouraged to review, having paper and pen in hand, the exercises proposed in the following sections.

1.2.2.1 Approximate Ratio between COD and VS

In many cases, it may turn interesting to compare the results of our own tests with those of the scientific literature. Unfortunately, there are no strict norms about the measure units in which the values must be expressed. One recurrent case is the use of COD or VS as different ways to estimate the potentially degradable organic matter or the amount of living microbes contained in a sample. From a formal point of view, there is no difference between adopting COD or VS as measure units of the organic matter concentration, since both are indirect indicators of the quantity of carbon contained in a given biomass. The important thing is to keep the coherence of the measure units when performing the calculations. Generally, COD is the preferred unit to measure the organic matter contained in liquid substrates (sludge, wastewater, whey, etc.) just because it is easier to perform a COD test on a liquid, while VS are easier to measure when analyzing solid biomasses.

It is possible to establish a quite approximated theoretical correlation between the said parameters, assuming that 50% of the VS mass is made of C (fairly true for most biomasses, especially vegetal ones, but not applicable to *all* substrates!).

From stoichiometry, it is known that 32 g of O_2 is necessary to completely oxidize 12 g of C.

Hence, the COD of a given quantity of VS is equal to:

$$COD = 0.5 \cdot SV \cdot \frac{32}{12} = 1.333 \cdot SV$$

Please note that VS can be expressed as % of the W.W. or as % of the D.W.

The convention adopted in this book expresses the VS as % of the W.W.

COD, in turn, is usually expressed in mg of O_2/l of slurry or liquid substrate (in rare cases also as g of O_2/g of DW).

Example No. 1

Calculate the COD of 60 g of sludge having 4% of VS

$$VS_{sludge} = 0.04 \cdot 60 \, g = 2.4 \, g$$

$$COD_{sludge} = 1.333 \cdot 2.4 \, g = 3.2 \, g \, \left(N.B., \text{these are absolute g of } O_2 \right)$$

$$COD_{sludge} = 3200 \, mg / 60 \, ml = 53.333 \, mg \text{ of COD}/l$$

Please note another convention, which is not a written norm but is routinely accepted worldwide—in general, it is assumed that sludge and liquid substrates have the same density of water; hence the 60 g of our sample is equivalent to 60 ml. In reality, sludge is denser than water—from 1020 to 1060 g/l—but this difference can be neglected in calculations for industrial purposes.

The correlation between COD and VS found earlier is not valid for all organic matters, as the next practical example shows.

Example No. 2

Calculate the COD of a 10 g sample of pure glucose.

The total oxidation of glucose is described by the following stoichiometric formula:

$$C_6H_{12}O_6 + 6O_2 \rightarrow 6CO_2 + 6H_2O$$

$$180 \text{ g} \quad 192 \text{ g} \quad 264 \text{ g} \quad 108 \text{g}$$

Hence, 180 g of VS (since glucose has no ash, i.e., it is fully volatile) is equivalent to 192 g of COD. Hence the equivalence ratio is 1 g COD = 1.0666 g of glucose.

By applying the same reasoning of the practical examples 1 and 2, it is possible to elaborate the conversion Table 1.6, which will turn quite useful when performing tests with pure substances employed as reference substrates.

TABLE 1.6

Equivalence Ratios between VS and COD of Some Pure Substances Employed for Performing Laboratory Tests in Biogas Plants

Substance	Chemical Formula	g VS/g COD
Vegetal and bacterial biomasses	≈50% C	1.3333
Glucose	$C_6H_{12}O_6$	1.0666
Sucrose (saccharose)	$C_{12}H_{24}O_{12}$	1.0666
Fructose	$C_6H_{12}O_6$	1.0666
Starch	$(C_6H_{12}O_6)_n$	1.0666
Cellulose	$(C_6H_{10}O_5)_n$	0.84375
Proteins	55% C + 7% H + 17% N + 21% O	≈0.42
Acetic acid	CH_3COOH	0.93
Sodium acetate	CH_3COONa	0.93
Propionic acid	$C_3H_6O_2$	0.6622
Sodium propionate	$C_3H_5O_2Na$	0.6622
Butyric acid	$C_4H_8O_2$	0.5525
Sodium butyrate	$C_4H_7O_2Na$	0.5525
Vegetable oils (assumed as oleic acid)	$C_{18}H_{34}O_2$	0.346

1.2.2.2 Methane Yield of a Substrate with a Known BMP

Generally, the scientific literature expresses the BMP of the substrates in terms of VS or COD. This is the correct mode to obtain universally comparable test results. The reason is that a given substrate may contain more or less moisture and more or less ash, depending on an infinity of factors. The substrate's BMP will be directly proportional to the VS percentage and should not diverge much from a characteristic figure, typical for the substrate in question. Many agricultural magazines and most suppliers of AD feedstock express the BMP either as a function of the W.W. or as a function of the D.W. Both criteria are somehow treacherous, because two different lots of a given substrate, e.g., silage, may have the same percentage of dry mass, but their VS percentage (and consequently their BMP) can be very different for several reasons. The following examples, taken from real cases, show how to perform the correct calculations in each case, so as to obtain coherent and comparable results.

Example No. 3

A source in the scientific literature states the following values for corn silage:

> TS = 30.7%
> VS = 95.5% (of TS)
> BMP = 310 Nl/kg VS

An agricultural magazine quotes the following data for the same substrate:

> TS = 30%
> VS = 91.5% (on TS)
> Biogas yield = 668 m³/ton TS

Is it possible to compare both sets of values?

Please note that the magazine adopts a wrong criterion, because it defines the yield in terms of biogas, but it does not define what percentage of CH_4 the biogas has. It is conventionally assumed that biogas is composed of 60% of CH_4 and 40% of CO_2. Under industrial plant conditions, or for some special pure substrates, the proportions of CH_4 and CO_2 are quite variable. Corn silage usually yields biogas with 55% of CH_4. Furthermore, the author of the article does not specify if the gas volume is expressed in Nm³ or Sm³.

Suppose, as first approach, that the author of the article has employed the most frequent conventions of the biogas industry. Replacing the TS by the usual equivalent VS content, we obtain:

$$\text{Biogas} = \frac{668\,[\text{Nm}^3]}{0.915\cdot 1\,[\text{t TS}]} = 730\,[\text{Nm}^3/\text{t VS}]$$

$$\text{BMP}_a = 0.6\cdot 730\,[\text{Nm}^3/\text{t VS}] = 438\,[\text{Nm}^3/\text{t VS}]$$

The BMP value thus obtained seems too high if compared to the typical BMP found in the scientific literature and is assumed as a *benchmark*. Let us recalculate the BMP assuming that the biogas contains only 55% of CH_4 (usual for substances composed mainly of carbohydrates, like the silage). Under such supposition, the BMP deduced from the magazine's article will be nearer to the value found in the scientific literature.

$$BMP_b = 0.55 \cdot 730 [Nm^3/t\ SV] = 401.5 [Nm^3/t\ SV]$$

If the author of the article normalized the volume in Sm^3, then the resulting BMP expressed in the correct units is:

$$BMP_c = \frac{0.55 \cdot 668 [Sm^3] \cdot 0.947 [Nm^3/Sm^3]}{0.951\ [t\ t.q.]} = 380 [Nm^3/t\ SV]$$

By reviewing the literature, it will appear that, depending on the authors of different studies, the BMP of corn silage varies from 338 Nm³/ton VS to 537 Nm³/ton VS. So, who is right?

We leave the reader to meditate on this apparent contradiction, since we will deal in detail the argument of the error propagation in the determination of the BMP of silage in the forthcoming chapters. The important concepts to keep in mind from this chapter are:

1. The BMP that can be effectively obtained from a given biomass in a given biogas plant is not a universal and absolute value.
2. Not all bibliographical sources are reliable.
3. It is very important to pay attention to the units in which the BMP is expressed, especially when the numerical values are very high, and even more especially when said values are published in "nonscientific" magazines, because quite often such articles are sponsored by companies or stakeholder groups having a particular interest in showing high productivity of their products.

1.2.2.3 Coherent Use of the TS, VS, and BMP Values

Example No. 4

Upon measuring the TS, VS, and BMP of a silage sample taken from our stock, we determine 31% TS, 29% VS (both expressed on W.W.), and 348 Nm³/ton VS. How many tons of silage should we feed to the digester, so as to bring the plant to work at its nominal power of 500 kW? The electric efficiency of the generator, according to the manufacturer's technical sheet, is 38%.

First, we need to calculate the primary energy necessary to feed the generator at full power during 24 h.

$$E = \frac{500 [kW] \cdot 24 [h]}{0.38} = 31{,}578.94 [kWh]$$

Given the LCV of methane, its necessary daily quantity, G, will be:

$$G = \frac{31,578.94\,[\text{kWh}]}{9.94\,[\text{kWh/Nm}^3]} = 3,177\,[\text{Nm}^3]$$

The total OL to feed the digester will then be equal to:

$$OL = \frac{3,177\,[\text{Nm}^3]}{348\,[\text{Nm}^3/\text{ton SV}]} = 9.13\,[\text{ton SV}]$$

Since the silage has 29% of VS on W.W., the total daily amount of silage to load on the plant's hopper, M, will be:

$$M = \frac{9.13\,[\text{ton SV}]}{0.29} = 31.5\,[\text{ton}]$$

N.B.: This is the typical (linear) calculation adopted by most biogas plant managers. Please note that this simple calculation assumes the total degradation of the silage (i.e., extracting its whole BMP) and a constant production of biogas to be directly proportional to the OL, which is not true in most of the cases.

1.2.2.4 Calculation of the HRT

Example No. 5

Consider the same data as in the former example no. 4. Assuming that the plant consists of a single digester with $4000\,\text{m}^3$ of useful volume and that, in order to feed it by means of a pump, the silage must be diluted with $4\,\text{m}^3$ of water per ton of fresh weight. What are the HRT and the SRT?

In this case, the HRT is coincident with the SRT, since we suppose that the digester is perfectly mixed. Such assumption is true for most of the existing plants, but it is not an absolute truth. We are also assuming that all solids will become biogas during the digestion process (not completely true, but acceptable supposition in the case of highly degradable matter like corn silage).

The total daily flow of liquid loaded to the digester, Q, results from the following formula:

$$Q = 31.5\,[\text{ton/day}] \cdot 4\,[\text{m}^3/\text{ton}] = 126\,[\text{m}^3/\text{day}]$$

By definition of HRT, we have:

$$\text{HRT} = \frac{V}{Q}$$

where

V = useful volume of the digester [m³]

Q = hydraulic load [m³/day]

Hence HRT = 4000/126 = 31 days

N.B.: Please note that the useful volume of the digester is never equal to the geometric volume, because a certain fraction of the total volume must be left free for the eventual formation of foam on the slurry's surface; hence it is impossible to fill a digester to its maximum (geometric) capacity.

1.2.2.5 Checking the OLR

Example No. 6

Calculate the OLR of the digester considered in the former examples.
By definition:

$$OLR = \frac{OL}{V}$$

Hence

OLR = 9130 kg VS/day/4000 m³ = 2.3 kg VS/m³·day

In a conventional biogas plant (stirred reactor, average feedstock solids content in the range of 9%–12%), the OLR should remain in the range of 2–3 kg VS/m³·day (it is advisable to keep it nearer to 2 than 3 for maximum efficiency of the conversion of biomass to methane). Hence, the digester is correctly loaded.

1.2.2.6 Optimizing the C/N Ratio of a Substrates Mixture

Example No. 7

In a farm with 100 pigs, the daily production of manure is equal to 1000 kg/day. Oats straw is available as cosubstrate. What is the quantity of straw that the farmer should add to the manure so as to keep the biogas plant working at its optimum capacity?
Data (see Table 1.4)

- C/N of swine manure = 4%/0.4% = 1000·(0.04/0.004) [kg/day]
- C/N of oat straw = 50%/0.5% = (0.5/0.005)·x [kg/day]
- Ideal C/N ratio of the mixture for optimum digestion = 150/5

The amount of straw, x, necessary to obtain a mixture with ideal C/N ratio is hence:

$$\frac{150}{5} = \frac{1000 \cdot 0.04 + x \cdot 0.50}{1000 \cdot 0.004 + x \cdot 0.005} = \frac{40 + x \cdot 0.50}{4 + x \cdot 0.005}$$

$$150 \cdot (4 + x \cdot 0.005) = 5 \cdot (40 + x \cdot 0.50)$$

$$600 + x \cdot 0.75 = 200 + x \cdot 2.50$$

$$600 - 200 = x \cdot (2.50 - 0.75)$$

$$x = 228.6 \,[\text{kg/day}]$$

1.3 Managing the Plant "by Tables": Limitations and Risks

Most industrial applications employ two different technologies of process control, called "open loop" and "closed loop" controls. We explain with a practical example the meaning and the advantages and limitations of each one, while quoting specific literature on the theory of systems control for the readers who will be willing to obtain a deeper and more specific knowledge on the argument.

Suppose we desire to heat a room with an electric stove. There are two ways to control the flux of energy to the stove:

1. By connecting the stove to the electric supply through a timer-controlled switch. In such a case, the timer turns on the stove at a predefined time and keeps it working during a certain time, in which the temperature of the room will reach a certain value. For sure, the final value will be higher than the initial one, but we cannot know it in advance, because it will depend on the border conditions (initial temperature of the room, thermal dispersions of the room's envelope, eventual voltage variations while the timer is *on*, etc.).

2. By installing a temperature sensor and a thermostat in the room, which will control a relay that turns the stove on and off. In such a system, the electric stove will heat the room until its temperature reaches the desired value set in the thermostat. The time to reach the said preset value will depend on many factors, the same as in the case (1), but the system (2) has the advantage of reaching exactly the temperature desired by the user, regardless of any other circumstance.

A system like the one described in the Example (1) is known as open loop control. We could well call it "dumb control system" because, by starting the timer, the electric heater will turn on regardless of the effective temperature of the room and will heat it during a time, predefined by the manufacturer with some general criterion, unknown to us, which not necessarily applies to our specific case. By adopting such a control system, most of the time the user will either waste energy by overheating the room (in cases when it is initially warm), or fail to reach a comfort condition during very cold weather, when the preset time may not be enough to bring the temperature to an acceptable value.

A heating control like the one described in the Example (2) is surely better, both from the energy consumption and comfort points of view, because it turns the heater on and keeps it on only until the desired temperature is reached.

Both control techniques are usually depicted in the technical literature with the diagrams shown in Figure 1.5.

In spite of an industrial biogas plant costing millions of euros, closed loop control systems are employed just for the temperature of the digesters and the motor unit, but the management of the biological process is mainly based on an open loop logic.

In practice, most biogas plant managers rely on tables of BMP, pH, VFA/TA (FOS/TAC), ORP, electric conductivity, and similar parameters, to decide how to manage their plants. The experience shows that the said tables are seldom reliable, because they come from different bibliographical sources, published by researchers who have performed measures in conditions that seldom apply fully to our plant, or our feedstock, or both.

The appealing concept underlying the use of tabulated values to manage a biogas plant is the linearity of the approach: the plant manager hopes to obtain the desired production of methane just by performing some simple calculations based on values that somebody has determined in advance. Such approach quite often results in producing less energy than expected while in other (rare) cases it may lead to methane overproduction, which means wasting both energy and feedstock. The most diffused approach in the biogas industry consists of maintaining the production at its nominal maximum but at the expense of sacrificing the efficiency in the conversion of the biomass

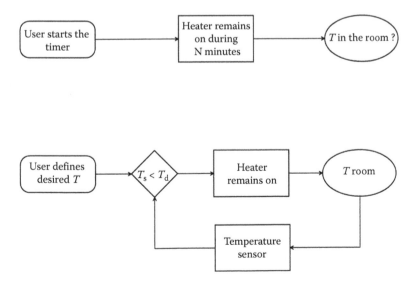

FIGURE 1.5
Block diagrams of an open loop control (above) and a closed loop control (below).

in methane. In any case, the economic result will be less than optimal. When underfeeding the plant, the biomass is digested with high efficiency but the energy production will be less than expected. In the opposite case, the nominal power of the plant is reached and the energy production kept constant, at the expense of consuming more feedstock than expected. From an environmental point of view, overfeeding results in emitting more greenhouse gases to the atmosphere (because of the incomplete digestion of the feedstock, the digestate will still have a residual methanogenic capacity) and requires to manage a bigger quantity of digestate. For these reasons, we can conclude that managing the biogas plant on the basis of just tabulated values is attractive in its apparent simplicity, but not effective and economically efficient.

Many biogas plant manufacturers propose control solutions to manage "automatically" the plant, apparently based on the closed loop concept, but it is necessary to point out that the theorems on which the closed loop control theory is funded are applicable only to *linear systems*, i.e., each action will produce a directly proportional reaction. Closed loop systems can of course control nonlinear processes, on condition that the control logic is also nonlinear, hence more complex. Unfortunately, the AD is a nonlinear process. We can even say that it is "complicatedly nonlinear." The mathematical function that better describes the degradation of the organic matter and its corresponding transformation to methane is indeed a complex exponential, as shown in the next examples.

1. Example of linear equation (for instance, the amount of natural gas delivered by a tube connected to a constant pressure grid during a certain time):

$$H = a \cdot t$$

    where

    H = total amount of gas cumulated in time t

    a = a proportionality coefficient depending on the grid's pressure and the tube's length and section

2. Mathematical model of the AD (aka the *Gompertz equation*), whose graphical representation is a curve called *sigmoid* because of its shape resembling an S:

$$H = a \cdot e^{\left[-b \cdot e^{-c \cdot t} \right]}$$

    where

    H = methane production cumulated in time t

    a = upper asymptote of the curve, physically corresponding to the substrate's BMP

    e = base of the natural logarithms = 2.71828…

b = a coefficient that is proportional to the acclimation time required by the bacteria (called *lag phase* in technical literature), physically meaning the necessary time to reach the maximum daily flow

c = a coefficient that is proportional to the maximum daily flow

t = elapsed time (the variable of the Gompertz equation, usually expressed in days)

Another form of the Gompertz equation usually found in the literature is the following:

$$H = a \cdot e^{-\left[\frac{c \cdot e}{a}(\lambda - t) + 1\right]}$$

where

H = methane production cumulated in time t

a = upper asymptote of the curve, physically corresponding to the substrate's BMP

e = base of the natural logarithms = 2.71828...

λ = lag phase (days)

c = maximum daily flow (Nml/day)

t = elapsed time (the variable of the Gompertz equation, usually expressed in days)

For the reader who is not familiar with exponential functions, it is interesting to graphically compare the proportional (linear) process (i.e., the assumed behavior when applying data from tables) and to compare it with the sigmoid curve, which is the real-life case (Figure 1.6).

When managing a biogas plant, it turns useful to know the total methane flow that a given quantity of biomass can yield, as a function of time. By definition, the flow is the quantity of gas produced in a unit of time, a quotient that can be calculated in each point of the cumulative volume curve by dividing the latter in very small intervals. In Mathematics, the said operation is called *to derivate*, and the curve thus obtained is called the *derivative*. The derivative of a sigmoid as a function of time (hence the curve representing the instant flow at any time) is a bell-shaped curve, represented by the following mathematical expression:

$$H' = a \cdot b \cdot c \cdot e^{-b \cdot e^{-c \cdot t}} \cdot e^{-c \cdot t}$$

Figure 1.7 shows very evidently the substantial difference between the "ideal" behavior that any biogas plant manager desires (linear degradation of the biomass, so as to obtain a constant biogas flow) and the real behavior of the AD of any substrate.

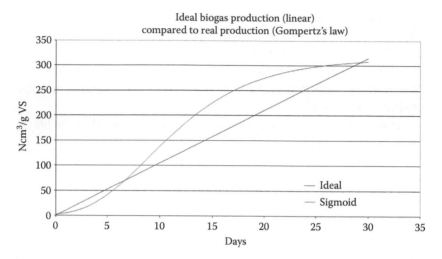

FIGURE 1.6
Cumulative methane production from the AD of organic matter: real biogas production compared to the assumption of linear production.

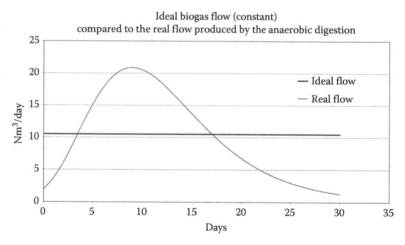

FIGURE 1.7
Daily biogas (or methane) flow as obtained from a batch test.

In practice, we need to apply the results of a batch test on a biomass sample (the sigmoid) to a plant that must produce continuously. How can we employ the result of the laboratory test to calculate the diet of our biogas plant? To answer this very practical question, we must consider that the final scope of all the tests described in advance is to calculate how much of a given feedstock must be loaded into the digester so as to obtain a given (average) methane flow. Flow is, by definition, the quotient of the gas volume (or the mass) by the unit of time. If we divide the sigmoid in intervals of 1 day, and tabulate the quantity of methane produced in each day, we will obtain the curve of daily gas flow as shown in Figure 1.7.

Fortunately, a continuous digester—i.e., a digester receiving a certain amount of feedstock each day—can be imagined as a battery of many batch digesters being fed sequentially, and hence each one will generate the same daily flow curve, but each curve would be delayed 1 day in comparison with its precedent, as shown in Figure 1.8.

Hence, under the given supposition, the total biogas flow produced by a continuous digester fed a certain daily amount of biomass, M, is equivalent to the sum of the instant flows of n batch digesters fed the same amount M, once every n days. The result of this operation, assuming that the plant is already working in steady state, is shown in Figure 1.9.

The former reasoning and the result shown in Figure 1.9, both explain three very important facts:

1. Now we can understand why most biogas plant constructors adopt the Continuously Stirred Reactor Tank (CSRT) technology. As can be seen from Figure 1.9, the said reactor type "linearizes" the behavior of a process that is not linear itself.

2. It should be clear now why in continuous biogas plants it is necessary to load a daily quantity of substrate slightly higher than the one calculated as the simple quotient between the daily necessary

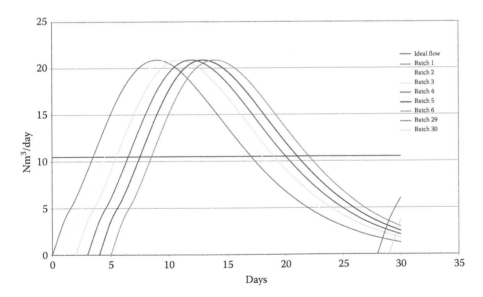

FIGURE 1.8
Simplified mathematical representation of a continuous digester as a group of sequentially fed batch digesters.

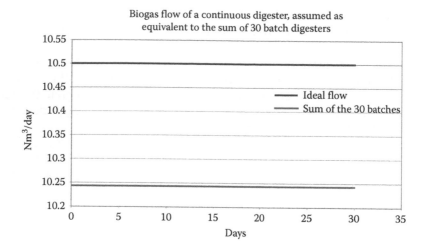

FIGURE 1.9
Real flow from a continuous digester with 30 days retention time, assumed as a group of 30 batch digesters, each one being loaded every 30 days with the same amount of feedstock that constitutes the daily ration of the continuous digester (violet line), compared to the flow calculated with the simple linear assumption, BMP/30 days (blue line).

methane quantity and the BMP. The reason is that the anaerobic degradation of any substrate has an asymptotic limit. Hence, in practice, it is impossible to extract 100% of the feedstock's BMP. The digester's efficiency (the effective amount of gas produced in a certain time compared to the BMP measured in batch tests in the same time) can reach high values, but it cannot reach 100% unless the HRT (or SRT) is long enough (theoretically infinite in the mathematical model of the Gompertz equation). A second reason, usually neglected by biogas plant designers and often forgotten by academicians, is that a digester is a microbial farm. As any other living beings, microorganisms grow and reproduce. Consequently, part of the feedstock contributes to increasing the mass of living bacteria in the sludge instead of becoming CH_4 and CO_2.

3. The constructors of biogas plants program their control systems on the basis of a "closed loop" logic, which relies on BMP values "from tables." The BMP is represented by the parameter a in the Gompertz formula, but the average biogas flow depends *also* on the coefficients b and c, which in general are not tabulated. Even in the case a biogas plant constructor should consider the values a, b, and c as variables for programming his control system, the said values depend not only on the digester's feedstock but also on the state of the bacterial system (in the following chapters we will define in detail what *state* means). Hence, the control systems sold by most biogas plant

constructors as "technologically advanced" (because based on a "closed loop" logic) result in being only partly effective. In practice, they contain a certain extent of the "open loop" logic, since their operation has been programmed in the factory on the basis of coefficients "from tables," and not necessarily coincident with those really describing the features of the feedstock and the activity of the bacterial consortium in a given moment.

1.3.1 Conclusions on the Use of Tables

The limitations of the biogas plant management arising from the use of tables are two:

(1) The tables only show *one parameter* of the whole AD curve, the BMP, which in turn is extremely variable as a function of several factors, like the chemical composition and grain size of the feedstock. (2) The other two parameters describing the AD dynamics are impossible to tabulate because they depend on the biodiversity and activity of the inoculum, the process temperature, the pH and alkalinity of the mixture under digestion, the reactor's geometry, the stirring intensity, the presence or absence of biocatalysts or inhibitors, and so on.

The problem is further complicated when the digester's diet is composed of mixtures of organic matter, since the BMP of a mixture is not necessarily the arithmetic sum of the individual BMPs. The explanation to this apparent paradox is that the anaerobic bacterial ecosystem thrives at its best when the C/N/P ratio is nearly 150/5/1. With this in mind, it appears evident that mixing two substrates, for instance both rich in N, will bring down the C/N ratio, creating in some cases a partial inhibition of the whole process, because of the defect of carbon and excess of ammonia. Conversely, a mixture of two substrates in such proportions that the resulting C/N ratio is nearly 30 could produce a synergic effect producing more methane than the arithmetic sum of the single BMPs measured separately.

A table of the BMPs of several common substrates is included just as an example.

Surely, the reader is already disagreeing with some of the said values, since they may differ too much from the values that can be deduced from the plant's operation. Feedstock suppliers and plant constructors may declare very different values too. Which value is then correct? The same perception that something was wrong when comparing BMP values of the same feedstock has driven the scientific community to question whether the test protocols for the BMP assay were wrong, given the high variability of the results found by different research groups.

Many experiments have been conducted to find out the reason of the big discrepancies between the BMP values measured by different laboratories. For instance, in 2011, Raposo et al. (2011b) conducted a study on 18 laboratories, who had to determine the BMP of a 100% degradable and pure substrate whose theoretical BMP is well defined—starch. Figure 1.10 shows the result.

The reader can now easily realize that, if the results are so extremely variable when measuring the BMP of a perfectly defined pure substrate, and if we discard the possibility of human errors (because the tests were performed by expert laboratory operators in controlled conditions), the causes of the variability of the results must be quite more elusive.

In the first approach, we can preliminarily conclude that the BMP does not depend only on the chemical composition of the substrate, but also on the bacteria that compose the inoculum, on the preparation of the sample, on the test conditions, etc., even the geometry of the reactors and their stirring have a certain influence, but of second order.

A joke well known in the scientific world, somehow similar to the famous "Murphy's law," is called "biological systems law":

> Under perfectly controlled laboratory conditions, any living organism will behave in a way completely different from the researcher's initial hypothesis.

Jokes apart, it is perfectly true and normal that in a given biogas plant, fed with the same biomass, with the same process parameters, and following the same routine, it is impossible to obtain a perfectly constant methane flow by just controlling the feedstock flow to the digester (unless the digester is

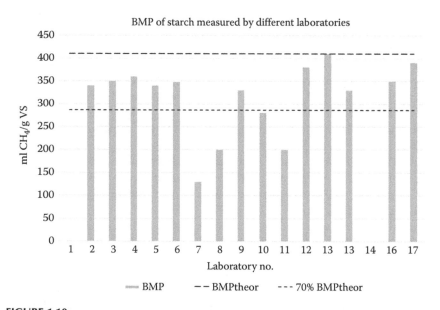

FIGURE 1.10

The BMP of starch, as measured by 18 different laboratories. (From Raposo, F., et al., *Journal of Chemical Technology and Biotechnology*, 86(8), 1088–1098, 2011b; graphics by the author.)

somehow overfed, as almost everybody does). Some sort of "closed loop control" (not necessarily an automatic device) must be applied. This objective fact brings us to the discussion in the next section.

1.4 The Dynamic Management of the Biogas Plant

A dynamic management approach is necessary to bring the biogas plant to run as smoothly as possible. This means that it will be necessary to monitor as many parameters as possible, among those that influence the three coefficients in the Gompertz sigmoid (a, b, and c), then check the flow and composition of the output biogas, and hence act consequently on the feedstock input. Unfortunately, no sensor exists (at least, not as commercial product on May 2017) that is instantaneously able to measure what is happening inside the digester, or to foresee what the methane yield of a given substrate will be in a given moment, without performing the necessary tests. The reader may have observed that even the same lot of feedstock from the same silo, in the same plant and under the same fermentation conditions, will yield different quantities of methane along time. The explanation of the said phenomenon is intuitive, because the ensilage process is in itself a kind of pre-fermentation of the vegetal biomass, but it continues in time, though at a slow speed. Hence, if we perform a BMP test the very month the biomass is ensiled, finding some given values for a, b, and c, after some months, it is sure that the said values will be different. Figures 1.11 and 1.12 show the thermal photos (taken with an infrared camera) of two trenches, both containing corn silage. The red spots show zones where the temperature is higher, clear signal that the silage is being oxidized by the aerobic bacteria into carbon dioxide (exothermal reaction, hence heat detected by the IR camera). As time passes and carbon is lost to the atmosphere, less carbon will remain in the silage, so a given quantity of feedstock will produce less methane. The causes of such uncontrolled oxidation are surely known to the reader—mice and crows pierce the membrane that covers the silage, allowing air to enter; the difficulty in closing the trench's front in an airtight way each day; more or less compaction and air trapped during the ensiling operations...

Several parameters have been proposed to monitor the AD process to "close the loop" of the control system with useful information, allowing us to dynamically vary the input of biomass to the digester, obtaining at any moment the maximum methane production with the minimum biomass consumption.

From a didactic point of view, it may turn useful to imagine the biogas plant as a car, which we need to drive for 1000 km at a constant speed of 150 km/h. In the system called "car," the instruments on the board panel and the view of the road through the front window provide the driver with the necessary information to "close the control loop," i.e., to press more or less the

FIGURE 1.11
IR photographs of a trench filled with corn silage. © Arch. Giovanna Barbaro, certified second-level IR thermography operator.

FIGURE 1.12
IR photographs of a trench filled with corn silage. © Arch. Giovanna Barbaro, certified second-level IR thermography operator.

gas pedal. Let us analyze some of the most common parameters composing our imaginary "board panel of the biogas plant," so as to understand their validity and why, even putting them altogether, it is impossible to keep the AD process at a constant rate.

1.4.1 The pH

It is well proven that the *Archaea* thrive at their best when their environment has a pH in the range of 6.8–7.2, as can be seen from the research by Andrews and Greaf (1971), as shown in Figure 1.13, taken from Gerber and Span (2008).

The fermentative bacteria, on their turn, prefer environments with pH ranging between 5 and 6.5, although in practice they can tolerate pH between 4 and 8.4. According to these facts, it appears evident that the optimization of one of the steps of the AD process will automatically hinder the correct development of the other, resulting in a lower specific production of methane for a given digester and feedstock. According to Mata-Alvarez (2002), the inhibition caused by NH_3 is predominant at pH > 7, while the inhibition caused by SH_2 predominates at pH < 7. The production of both inhibiting compounds together with CH_4 is unavoidable with some substrates (please refer to Table 1.7). Hence, in extreme conditions, it is possible that both compounds reach toxic levels, causing the biologic collapse, even if the pH remains very near to 7.

In the concrete case of digesters fed with manure, its high buffer capacity (aka alkalinity) prevents the pH from changing appreciably. When the pH eventually reaches an alarm threshold, it is because the concentrations of SH^2 or NH^3 have become so high, that it is usually too late to save the

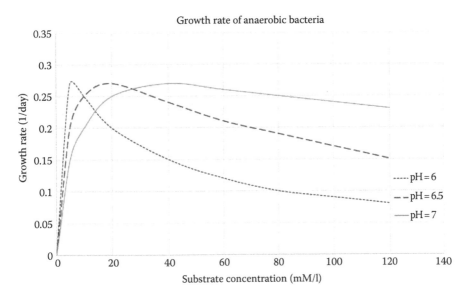

FIGURE 1.13
The reproduction speed of the *Archea* as a function of pH and substrate concentration. (From Andrews and Greaf, 1971; quoted by Gerber, M., Span, R., *International Gas Union Research Conference*, Paris, France, 2008; graphics by the author.)

TABLE 1.7

Indicative BMPs of Several Agricultural Substrates

Substrate	Average BMP (Nl CH$_4$/kg SV)
Swine manure	450
Cattle manure	250
Horse manure	460
Sheep manure	200
Manure from stable (with straw)	225
Chicken dung	450
Corn straw	410
Corn silage	550
Rice straw	220
Grass	410
Vegetal waste	350
Sewage water	420

digester from the biological collapse. We can conclude that pH is not a 100% reliable control parameter. In our analogy with the car's instruments board, we could compare monitoring the pH with the oil pressure warning: it will light on when the motor is nearly, if not already, seized.

Measuring the pH has the only advantages of being cheap and easy.

1.4.2 The Relationship between pH and Alkalinity

On the basis of the former considerations, if pH and alkalinity are measured (i.e., the resistance of the systems to pH variations), then we could judge with higher accuracy the stability margin of the biological process. We already explained that the *Archaea* thrive if the pH remains between 6.8 and 7.2. From similar tests, it is known that the ideal alkalinity must be in the range of 8000–9000 mg equivalent of calcium carbonate per liter (mg eq. CaCO$_3$/l), since higher values may start inhibition and dissolution of the active granules, while lower alkalinity values will allow wider pH excursions.

In the AD process, the alkalinity is given by the presence of carbonates and bicarbonates (dissolved carbon dioxide), ammonia, phosphates, silicates, VFA, and the total alkalinity is usually high (in the order of several g equivalent of carbonate/l of slurry).

The constructors of biogas plant that employ the alkalinity adjustment as control method (only a few, since this is an obsolete technique), install automatic dosing systems of sodium carbonate or bicarbonate connected to a system that measures the pH and the alkalinity. For sure the reaction of the VFA with the carbonate, or bicarbonate, immediately neutralizes the pH when this latter tends to drop, though there are some drawbacks that must be accounted for:

1. The reaction of the carbonate or bicarbonate with the acetic acid produces sodium acetate (which is exactly what the *Archaea* need), but the alkalinity also reacts with the SH_2 (always present in biogas plant running on silage and manure). Both reactions will immediately free CO_2, hence the percentage of methane in the biogas will drop.

2. The sodium remains in the digestate at the end of the process. The most common use of digestate is as fertilizer, but the accumulation of sodium in the ground could cause the opposite effect: the salinization and consequent aridness of the soil.

3. The pH is usually measured continuously with a simple electrode, but the electrode requires an automatic cleaning system to avoid fouling, which is not at all simple!. The measured pH values may vary, depending on the position of the electrode within the digester. On the other side, the alkalinity can be measured only with titration or spectrophotometric methods, which are difficult to implement for automatic continuous measures.

4. The method of the double measure, pH and alkalinity, is extremely reductive and of "linear" conception. Its validity is doubtful in spite of being much diffused in Central Europe and Italy, where it is known with the German acronym FOS/TAC. The installation cost of the automatic dosing devices for the alkaline solution, summed to the cost of the reagents, not necessarily, will be compensated by the eventual benefit. When the alkalinity falls, it is more advisable to reduce the substrate's input flow and/or to increase the percentage of manure in the total feedstock.

5. A mixture of magnesium hydroxide and calcium hydroxide (spent lime), or even ash from the combustion of vegetable biomass (mainly composed of potassium hydroxide), is a good or even better additive than sodium bicarbonate and more environmentally friendly for the soil that will be fertilized with the digestate.

1.4.3 The Chemical Composition of the Substrate

This technique is another example of "management with tables." Theoretically, the content of C, H, O, and N can be directly correlated with the BMP of a given substrate. This is more or less valid for pure, fully degradable substrates having a grain size less than 1 mm. Unfortunately, the elementary chemical composition of biomass cannot be correlated with the speed at which methane will be produced, nor does it give any idea of how degradable the biomass will be. Several other factors will influence the BMP, such as, the composition and the metabolic activity of the bacterial ecosystem in the inoculum, the grain size of the substrate, any eventual pretreatment the substrate may have undergone that alters the degradability of the more complex molecules, etc., a full list would be too long to include here.

Some laboratories active in the biogas market calculate the BMP on the basis of chemical composition instead of measuring it directly. The reasons, in the author's opinion, are most probably the result of marketing interests rather than of technical considerations. The analysis of the substrate's elementary composition requires specialized instruments and qualified personnel. Hence, it remains an almost impossible task for the biogas plant manager, ensuring thus the continuous earnings to the laboratories who will benefit of "captive customers" as long as the biogas plant is active. In practice, since the BMP calculated on the basis of chemical analysis corresponds to the theoretical maximum, the results are unrealistic. The biogas plant manager will be told about high methane yields from a given industrial substrate, or of a given energy crop cultivar, so as to convince him to buy the substrate or the seeds.

When formulating the diet of the digester on the basis of the chemical analysis of the substrate and theoretical formulas, the results will be unreliable. As can be seen in Figure 1.10, even when digesting a pure and fully degradable substrate like starch, the real methane yield can be much different from the theoretical BMP value. As a rule, it is advisable to distrust BMP tables redacted on the basis of the chemical composition of silage, fodder, or other substrates. In some rare cases, the activity of the digester's bacterial ecosystem could be enough to reach the theoretical BMP under laboratory conditions, but this does not guaranty the complete degradation of the substrate under real plant operational conditions. The explanation is simple: if we remember the Gompertz equation, already explained in Section 1.3 and shown in Figure 1.6, it appears evident that the BMP represents only the coefficient a in the said equation. The BMP tables supplied by animal fodder producers and by producers of "special cultivars for energy crops" never show the coefficients b and c of the Gompertz equation, for the simple reason that the chemical analysis method cannot calculate them.

As an example, let us analyze a real case on the unpleasant consequences of taking table values as absolute truths. It happened in Italy that a biomass sales representative and his client, owner of a biogas plant, had both read in a specialized agricultural magazine that fatty matters are characterized by particularly high BMP values. Knowing that the byproducts from the oil industry always contain a certain amount of residual fats, they superficially concluded that the said type of residual biomass should be optimum for AD. In fact, the magazine in question did not include any complementary information that could have helped in defining a strategy for employing the said oily feedstock in the biogas plant. Indeed, the article defined neither the necessary time for the total conversion of fats into methane, nor the maximum gas flow that could be attainable, nor the necessary time to reach the said maximum flow. In other words: only the coefficient *a* of the Gompertz equation was known, but the coefficients *b* and *c* were not. Apart from being difficult to degrade and sometimes even being toxic for some bacterial ecosystems, in practice fatty substances require *always* 50–60 days for their complete digestion. Hence, in spite of the high theoretical BMP resulting from chemical analysis, fatty substrates will never be able to produce relevant quantities of methane in biogas plants characterized by HRT shorter than 45 days. In our study case, the result was catastrophic: the bacterial ecosystem was not specific enough to be able to digest fats, so the biogas plant suffered a production block for several days.

1.4.4 The VFA Profile

This one is theoretically quite an accurate technique but in practice, it is difficult, expensive, and too slow to apply. If we could know at any time the concentration of acetic, butyric, and propionic acids, (some authors include valeric acid too), we could (based on tables from laboratory tests!) infer what the state of the AD process is. The practical problems of analyzing the profile of the VFA are multiple:

1. The chemical analysis methods (titration, fractioned distillation) of very heterogeneous volatile fatty acid mixtures characterized by low concentrations (in the order of a few ml VFA/l of slurry) are not very selective, just because the VFA are chemically very similar.

2. Chromatographic methods (gas and liquid) are quite suitable because they are selective, but require expensive instruments and the intervention of technically qualified operators. For sure, a biogas plant manager who must simultaneously take care of the farming activities, of the biogas power plant, of the feedstock supplies, and so on, cannot buy a chromatographer and analyze the sludge every day.

3. To be useful for the optimization of the biological process, the determination of the VFA profile should be carried on in real time, and the information thus obtained should be the input of some kind of "closed loop" control system, capable of controlling with nonlinear logic the stirring, the feedstock input, the pumps, the power plant, etc. Unfortunately, no sensor exists that is capable of measuring, continuously and selectively, the VFA's concentration profile.

1.4.5 Monitoring the Biogas' Flow and Composition

Both the instantaneous gas flow and the biogas composition depend on a combination of three factors: the substrate's chemical composition, the inoculum's bacterial activity (depending in turn on external factors like temperature, pH, concentration of inhibitors like ammonia, etc.), and the ratio between the living mass of the bacterial ecosystem and the mass of feedstock to be digested. The reproduction and metabolic rates of the different species of *Bacteria* and *Archaea* composing the ecosystem of the inoculum are diametrically opposite—under mesophilic conditions, the archaea reproduce themselves more slowly than the bacteria, while under thermophilic conditions, the biodiversity of the fermentative bacteria is smaller. For these reasons, the proportions of hydrogen, carbon dioxide, and methane in biogas depend on the state of the digester. In this context, *state* means "a set of several biological variables" (respiratory rate, reproductive rate, concentration of the different nutrients, etc.). It could be possible to better control the AD process by establishing in the laboratory the correlation between such factors like pH, alkalinity, BMP, OLR, VFA concentration profile, concentration of dissolved H_2, and the flow and composition of biogas. If monitoring all those parameters in real time was possible, it could be then possible to employ *several tables* linked by means of digital controllers programmed with *fuzzy logic* (a special technology of control that, unlike digital logics that only admit 1 or 0, white or black, true or false, admits intermediate values like "partially true," "light gray," etc.). Unfortunately, although the chemical composition of biogas is relatively easy to monitor in real time, analogously to the pH and the other parameters already examined, it is not reliable as single criterion for the biogas plant's management, since it is not directly correlatable to the state of the microbial ecosystem in the digester.

1.4.6 Monitoring the ORP

The measure of the ORP is easy to perform by means of electric probes, similar to those employed for measuring the pH. As in the case of the latter, the ORP measures the *intensity* of the reactions rather than the concentration of the reagents. Since a (theoretical!) correlation between the production of acetic, propionic, and butyric acid with the ORP exists, by measuring the latter it could be possible to know the stability degree of the AD process. It is well known that values in the range of –220 to –290 mV are good indicators of the "good health" of the process (equilibrium between the production of acetic acid and the intermediate products, propionic and butyric). ORP values within –280 and –350 mV favor the production of propionic acid and hence indicate the beginning of the inhibition of the methanogenesis. On the other hand, ORP values in the range of 0 to –150 mV indicate that propionic acid is being produced, regardless of the pH and the alkalinity, minimizing the production of acetic acid and hence of methane. If O_2 is bubbled in the digestion broth, so that the ORP can raise from –230 to –180 mV, the levels of SH_2 will drastically drop, but at the same time, the methanogenesis is inhibited. Consequently, the ORP should be maintained stably in a restricted range, within –230 and –220 mV, a task that is not at all easy from a technical point of view.

The main disadvantage of the ORP as control parameter of the AD process is the scarce reliability of its continuous measure, because the sensors get dirty and provide wrong signals. Another technical problem is

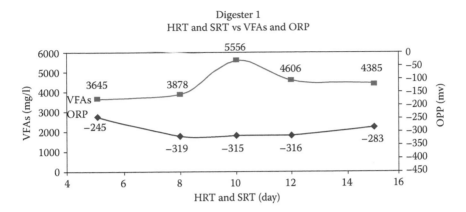

FIGURE 1.14
Concentration of VFA at different HRT, compared to the ORP values, according to Lee Sung (2008). (Courtesy of the University of Canterbury, Christchurch, New Zealand.)

to define where to place the sensor: in the laboratory it is easy to place the ORP electrode in the center of the reactor, but in the case of a digester having several thousands of cubic meters, the value measured in a given point may be different from the one measured in another point. Furthermore, ORP sensors usually have slow response times when working in anaerobic conditions—some probes may require from 8 to 24 h to stabilize their readings, a time which is long enough for undesired ORP variations to happen, which would not be detectable. Finally, ORP probes immersed in chemically aggressive environments, like digestion broth, require frequent calibrations.

The correlation between ORP and VFA concentration is not very good, as demonstrated by the experiments of Lee Sung (2008), shown in Figure 1.14.

1.4.7 Monitoring the Electric Conductivity

The electric conductivity is a parameter that can be checked very easily with a low-cost instrument called conductivity meter. The method's validity will be discussed in detail in Chapter 6, Section 6.2.1. In the context of the present chapter, we can simply state that it is the umpteenth table of values that allegedly allows a straightforward (i.e., linear) control of the biogas plant. The general idea is that the efficiency of the AD is linearly proportional to the electrical conductivity of the inoculum, as shown in Table 1.8.

The argument against such method is that microorganisms have a high capacity to adapt to hostile environments—like high salinity sludge—a fact experimentally verified by the same researchers in a real-scale biogas plant. Hence, Table 1.8 is to be considered as a merely indicative guideline. The author of the present book has successfully tested marine biomasses and *mozzarella* cheese whey having high salinity (about 40 mS/cm), observing

TABLE 1.8

Correlation between Electric Conductivity and Net Methane
Yield at 38°C

Electric Conductivity [mS/cm]	Net Methane Yield [%]
<12	100
20	85
30	65
40	32
>50	0

Source: Derived from the Curve Published by Garuti, M., et al.,
Conducibilita elettrica, utile monitorarla nel digestato,
L'Informatore Agrario, 40, 2014.

a certain degree of inhibition in comparison with samples containing less
salinity, but without noticing such a linear and steep reduction of the BMP
as the one found by the quoted researchers. It must be pointed out that a
digestion process can be collapsed in spite of the inoculum's salinity being at
acceptable levels, and hence measuring the conductivity does not provide a
reliable diagnostic in case of problems.

The main advantages of the electric conductivity method are its low cost
(<100 € for a pocket conductivity meter), its speed, and its easiness to perform
the measure. The big disadvantage is its lack of selectivity: One cannot know
if a high conductivity value eventually encountered in a biogas plant is the
result of a high concentration of salts or of ammonium ions. Furthermore,
the electric conductivity rises proportionally to the temperature; hence
Table 1.8 may be valid for mesophilic digesters, but it cannot be extrapolated
to thermophilic plants.

With the limitations of the method in mind, it may turn useful checking
the electric conductivity at least once a week in biogas plants fed mainly
with chicken dung, or in those plants where the solid feedstock is diluted
with the liquid fraction of the digestate. In the latter case, the increase of the
salts concentration with time is unavoidable; hence, it is advisable to employ
just water (better rain water) to dilute the feedstock. Another reason for the
salinization may be the substrate itself: biomasses cultivated on high salinity
soils will certainly contain more salts (ash) that will dissolve and remain in
the sludge once the organic matter is degraded. Another case when checking the electric conductivity may turn useful is that of plants where caustic
soda, sodium carbonate, or bicarbonate are systematically employed to control the alkalinity. According to the quoted authors, the best solution to add
alkalinity to the sludge without loading it with much sodium (i.e., keeping
the conductivity within acceptable levels without consuming water for the
dilution) consists of replacing the liquid fraction of the digestate with swine
or bovine stall effluents. Nevertheless, in the latter case the concentration of
ammonium ions will increase, with potentially toxic effects on the process.

1.5 The Outsourced Biological Management Service: Limitations of the Traditional Techniques

Farmers usually engage veterinaries and agronomists for the formulation of their animals' diet, and so the idea of hiring an external consultant, or paying the constructor of the plant for managing "the bacterial farm," results in being culturally acceptable for them. The drawback is that such *outsourced management* is more expensive, and implicitly excludes the permanent surveillance of the biological process. To remain competitive, the laboratories that offer biological management services limit their intervention to one weekly visit, or even a visit every 2 weeks, taking some samples that are analyzed according to precise routines (in general, in a laboratory hundreds of km away from the biogas plant). The result is the biogas plant manager receiving a report with the outcomes of the visit several days later, when such information is no longer useful. Quite often, the biogas plant manager does not know what to do with a report that, in general, does not contain practical indications about what to do on the basis of the laboratory's results, and hence he just archives the report. In other cases, the report only indicates that the biogas plant manager should add an integrator, the so-called "magical powders," punctually supplied by the plant constructor or by the laboratory itself. As an example, Figure 1.15 shows a laboratory test report, in which for privacy reasons the customer's and supplier's names were cancelled.

In this case, all the tests performed by the external laboratory could have been easily carried out on site by the biogas plant manager, as they require simple and cheap instruments (apart from the FOS/TAC, German name for the VFA/TA test, which will be discussed later). The fundamental advantage of performing the tests "at home" consists in having the results immediately available to act immediately if necessary. Hence, it does not make much sense to pay an external laboratory for getting delayed information

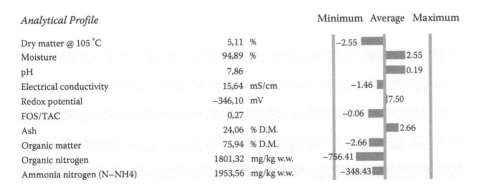

Analytical Profile			Minimum	Average	Maximum
Dry matter @ 105 °C	5,11	%	−2.55		
Moisture	94,89	%			2.55
pH	7,86				0.19
Electrical conductivity	15,64	mS/cm	−1.46		
Redox potential	−346,10	mV		7.50	
FOS/TAC	0,27		−0.06		
Ash	24,06	% D.M.			2.66
Organic matter	75,94	% D.M.	−2.66		
Organic nitrogen	1801,32	mg/kg w.w.	−756.41		
Ammonia nitrogen (N–NH4)	1953,56	mg/kg w.w.	−348.43		

FIGURE 1.15
Example of a report about chemical analysis performed on a sample taken from a digester.

Analisi

COA n°	███████	Ordine n°	███████	Data	14.02.2013

Sample no.	███████
Date of receipt	13.02.2013
Designation	Bioreattore
Sample type	Fermenting samples
Sampling date	07.02.2013
Sampling by	Customer
Sample amount	ca. 100 ml
Sample container	Container in PE with stabilizer
Number of containers	1
Start of analysis	13.02.2013
End of analysis	14.02.2013

Physical tests

Campione n°			13-018478-01
Sigla Cliente del Campione			Bioreattore
Dry matter	% w/w	OS	6.8
Organic dry matter (550°C)	%	OS	5.1
pH-value		W/E	7.5
Loss on ignition (550°C)	% w/w	DM	74.3

COA n°	███████	Ordine n°	███████	Data	14.02.2013

Short-chain aliphatic acids

Campione n°			13-018478-01
Sigla Cliente del Campione			Bioreattore
Acetic acid	mg/kg	OS	500
Propionic acid	mg/kg	OS	140
n-Butyric acid	mg/kg	OS	170
Isobutyric acid	mg/kg	OS	<50
Valerianic acid	mg/kg	OS	<50
Isovaleric acid	mg/kg	OS	<50
Caproic acid	mg/kg	OS	<50
Acetic acid equivalent	mg/kg	OS	729
VOA (titr.)	mg/kg	OS	3.900
Buffer capacity	mg/kg	OS	13.000
VOA (titr.) / TAC		OS	0.31

Nutritional values (total contents)

Campione n°			13-018478-01
Sigla Cliente del Campione			Bioreattore
Ammonium (NH4)	kg/t	OS	2.79
Ammonium-nitrogen (NH4-N)	kg/t	OS	2.2

FIGURE 1.16
Another example of laboratory report on samples taken from a digester.

in change. In the example above, the usefulness of the results provided by the laboratory is also questionable. As discussed before, the pH serves to almost nothing. The percentages of DM, ash, and moisture (moisture is just 100%−DM%, hence including it adds no useful information to the report), are not very useful without a context. ORP and electrical conductivity, like pH, are of little or no use to optimize the overall efficiency of the biogas plant, although some laboratories and plant constructors exaggerate much their importance. The laboratory of this example provides some more information than others do, since it indicates how the measured values compare to average indicative values, but does not provide any useful indication on what to do for reaching the recommended values.

Figure 1.16 shows the results, for the same plant and more or less at the same time, of the tests carried out by another competing laboratory. In this case, the tests are more complete and include some parameters that are useful for a correct management like, for instance, the profile of the VFA, but also in this case, the laboratory's report does not include any comment or recommendation on what the biogas plant manager must do. Once again, the shown values were measured 1 week after taking the sample and sent to the customer the following day, too late. Physical tests could have been spared, since the biogas plant manager can easily perform them *in situ*, as will be shown in the forthcoming chapters, and are useful only if the results are evaluated and actions are taken immediately, not 1 week later.

1.5.1 Conclusions

Hiring external consultants is not a wrong idea in itself, but it is necessary to choose the more professional ones. Beware of those that, apart from performing the tests, provide boosters and similar products (since quite improbably will they be impartial), and last but not the least, it is essential that the biogas plant manager gets the results of the tests within 24h from the sampling, including indications on what to do according to the measured values. External consultants or laboratories are indeed necessary to perform those tests that the biogas plant manager cannot easily carry out in the field (for instance, the profile of the VFA). The biogas plant manager should perform *in situ* the simpler tests, whose simplicity does not hamper their importance for an optimal management of both the biogas plant and the resources.

1.6 The Automatic Titrator: Myths and Legends

From a formal point of view, the use of the automatic titrator should be included among the different techniques discussed in Section 1.4. The author considered that it was worth dedicating a separate section for this argument, since in some countries—mainly Germany, Austria, and Italy—the usefulness of the said technique has been exaggerated, more because of marketing than because of scientific evidence. The blind and automatic application of a simple titration method (a common practice in Chemistry) is so banal and "linear" that it easily captures the fondness of the biogas plant managers and the trust of the investors and banks that finance biogas plant projects. The current belief that a titrator is reliable, just because it is automatic and hence "fool proof," leads to the misconception that employing it as a main diagnostic instrument will result in a reliable operation of the biogas plant, according to a linear logic based on tables. Let us start our analysis of this technique, called FOS/TAC or VFA/TA, from some theoretical concepts. FOS stands for *Flüchtige Organische Säuren*, which means volatile fatty acids in German. The final products of the organic matter's degradation are acetic, propionic, and butyric acids, called *volatile* because of their low molecular weight. They evaporate easily and cause the typical smell of rancidity. We already know that the precursor of methane is the acetic acid (or, more properly, its salts: sodium, potassium, and calcium acetate). Propionic and butyric acids, if present under certain concentrations, are anyway converted into acetic acid and the methanogenesis process continues undisturbed. If the digester is overfed, or if for any reason the intermediate stages of the fermentation of "heavy" biomasses (rich in proteins and fats) get blocked, then butyric and propionic acids begin to accumulate, until they consume all the alkalinity and end up inhibiting the archaea by lowering the pH. On the other hand, TAC stands for *Totales Anorganisches Carbonat* (total inorganic carbonates, aka total alkalinity). In practice, the presence of carbonates produces an effect of resistance to the pH changes, called "buffer capacity" or just alkalinity. Alkalinity is relatively easy to measure with automatic instruments called titrators. The titrator can be employed to measure the *total quantity* of VFA too, which is usually expressed *as if* they were *pure* acetic acid. We highlight "total quantity" and "as if," because what is actually present in the fermentation broth is a mixture of propionic, butyric, and acetic acids, and the exact percentage of each is unknown. According to the advocates of this technique, the VFA/TA ratio indicates the health state of the AD process. Below a certain value (i.e., low concentration of VFA), the bacteria are undernourished and the digester should be fed with more substrate. Over a certain value, the quantity of accumulated VFA risks consuming all the buffer capacity of the carbonates, which means that either the plant is overfed, or that the population of fermentative bacteria is about to overwhelm that of the archaea. In other words, it means that the latter are not able to consume all the available acetic acid, or that the penultimate link of the trophic chain, the conversion of butyric and propionic acids into acetic acid, is missing. Returning to the analogy with the instrument panel of a car: imagine reading the indication of the speedometer and the tachometer and then calculating the quotient. Theoretically, the driver would understand if the motor is running at the correct conditions for an optimum economy of fuel just by comparing the said quotient with a table defined by the car manufacturer. In practice, such approach is too reductive (once more, an example of the engineers' will to reduce complex phenomena to tables and linear equations!). Let us see what the weak points of the VFA/TA method are. Table 1.9 shows the recommended values for the VFA/TA ratio.

TABLE 1.9

The Original FOS/TAC Table According to Lossie and Pütz (2008)

FOS/TAC Ratio	Indication	Actions to be Taken
>0.6	Excessive organic load	Stop feeding the digester
0.5–0.6	High organic load	Reduce feedstock input
0.4–0.5	The digester is at the limit	Monitor the digester carefully
0.3–0.4	Ideal conditions for the production of biogas	Keep feedstock input constant
0.2–0.3	Insufficient organic load	Increase gradually the feedstock input
<0.2	Extremely low organic load	Increase quickly the feedstock input

From the explanations given, it is evident that the use of this technique can be included among the management "by tables." The FOS/TAC method is based on tests performed by some researchers under certain conditions *that not necessarily correspond to the ones of our plant*. Then somebody has redacted a table that indicates to add more or less feedstock to the digester, assuming that the said feedstock is the same employed as a benchmark for producing the table, and without any relationship with the overall conversion efficiency into methane of the whole process. In our analogy with a car's instrument panel, the ratio between the readings of the tachometer and the speedometer could be tabulated as an indication of the optimum set point for the motor, when the vehicle runs on a flat road, with its gear in fifth position and the motor fed with a well-defined fuel. The said table would be useless if the vehicle was up or down a slope, or its gear was in another position, and the tabulated value does not give an indication of whether the fuel tank contains petrol 98, petrol 95, kerosene, or red wine. To employ such a logic with success, a table for each possible driving condition would be required. Although nobody would ever fill the tank of his car with a fuel different from that established by the car's manufacturer, the same does not apply to anaerobic digesters, which by their own nature should digest any organic feedstock or mixtures of different biomasses, depending on their availability and economic factors. We can then conclude that a table redacted in the way described by the titrator's manufacturer—based on the operation of a single CSRT digester fed with bovine manure and corn silage—could be valid for the management of a similar plant with the same substrates. Such is the case of 98% of the agricultural biogas plants in Central Europe and most of the plants built in Italy between 2004 and 2010, but it need not necessarily be applied to a plant running on municipal wastewater sludge, or installed in other countries where corn is not the main feedstock for AD. Even in the case of biogas plant fed with corn silage and manure, we have seen that the degradability and hence the methane yield of the first is very variable and the features of the second are variable too. Manure is different depending on the age of the animals, the quantity and quality of fodder they are fed, its storage conditions and time, and dozens of other factors that would be too long to list here. Furthermore, if the plant had two or more digesters connected in series, or if it was of the *plug flow* type (the so-called "sausage-type digesters"), it would be necessary to produce an independent FOS/TAC table for each case. If the plant's diet changes from silage to other substrates, for sure the blind application of Table 1.9 would be, to say the least, risky.

A second problem with the VFA/TA method is its absolute lack of selectivity. Remember that the VFA value is measured *as if* it was 100% acetic acid. In practice, if only propionic and butyric acids, together with a certain alkalinity were present, the VFA/TA ratio could be good or acceptable according to the table, but the digester would be already under biological collapse. The *said condition is not a theoretical speculation, it is relatively frequent in plants fed with silage and the author has found it in some occasions*. The case of the collapsed digester described before can be explained with an analogy: imagine a clinical test that has been developed by statistically measuring the ratio between cholesterol and glycemia on a population of vegetarian and sportive persons, assumed as the paradigm of "healthy" people. If the ratio falls within a certain range, the test diagnoses the person as "healthy." Now imagine a sedentary person who is a lover of salami and sweets. It is not difficult to foresee that in such a case the ratio between cholesterol and glycemia can easily fall within the range of values assumed as "healthy," since the test considers only the ratio and not the individual values; nevertheless, the patient in question would risk both a heart attack and diabetes.

Contrary to the "folklore" of the biogas industry in many European countries, the VFA/TA is neither a modern method, nor is it a German invention. According to D. Bolzonella, the method was first published by Ripley et al. (1986) and its original name was "partial to total alkalinity ratio." The difference between Ripley's method and the variant called FOS/TAC by the German biogas industry is only in the final pH values assumed as the titration limits.

Why is then the FOS/TAC method so diffused in Italy, Romania, Austria, and Germany, while in other countries like Sweden, England, Spain, or France it is seldom employed? It is not by chance that the FOS/TAC method is diffused in countries where biogas plants are subsidized and hence many investors turned to this industry in search of stable profits. Since banks and venture capitals pretend any possible guaranty, keeping a register of the VFA/TA values measured from time to time serves to keep them happy. Perhaps because pushing a button, reading a value in an instrument's display, checking a table and finding that everything is OK sounds very easy and "linear." In most cases, the process may be running correctly, but if problems are lurking, the method does not see them. It must be said that a biogas plant fed with cow manure and silage is quite unlikely to suffer biological collapse, hence taking a VFA/TA reading once a week or not taking it at all, makes no difference in improving the digestion efficiency. Furthermore, even the daily measure of the VFA/TA does not prevent the biological collapse that is always possible when a feedstock containing anti-nutrients, like almost all cereals, is the only substrate fed to the digester. Again, such a condition is frequent in Central Europe, Italy, and Romania, where cereal silage is the main feedstock for biogas plants, but it is surely not the case in the rest of the world. Furthermore, the use of a titrator as sole instrument for the biogas plant's control makes impossible to diagnose in advance the formation of hydraulic short circuits (a problem described in detail in Section 4.1.3, one of the most sneakiest problems always lurking in the biogas plant).

As a demonstration of the limitations of the VFA/TA method, we will present the case study of a biogas plant in Northern Italy, 1 MW electrical power, fed exclusively with corn and triticale silage and (in small proportion) bovine manure. Figure 1.17 shows that from day 18 to 23 the VFA/TA ratio was kept at 0.25 (i.e., very close to the ideal value according to the advocates of this method). However, the energy production is clearly unstable, shifting from the maximum to the minimum peak in the period under examination. Such an anomalous behavior can be explained because in that moment, the total amount of the VFA was almost all butyric and propionic acids, and the methanogenic phase of the AD process was already inhibited. According to the criterion established by the manufacturer of the FOS/TAC titrator, already shown in Table 1.9, a value equal to 0.25 means that the plant should receive its nominal daily ration (about 50 tons/day of silage). Nevertheless, in our case study a reduced ration (about 12 tons/day of silage) fed from day 23 was already enough to cause the FOS/TAC value to grow almost exponentially, while the energy production continues to follow an erratic pattern.

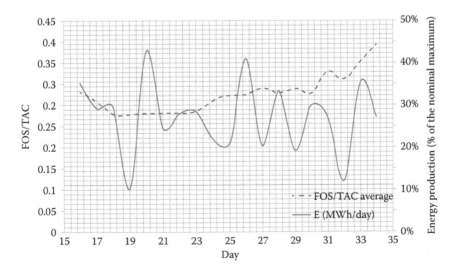

FIGURE 1.17
Comparison of the energy produced by the biogas plant and the VFA/TA parameter, measured with a FOS/TAC titrator.

In this case, the FOS/TAC test was only useful to confirm what we already knew—that the plant had biologically collapsed.

To employ the FOS/TAC titrator correctly, it is necessary to respect some practical rules. First, be sure of performing correctly the test. It may sound obvious, but the author has found many people measuring the necessary amount of distilled water with a glass without graduations, or not using correctly the automatic pipette. It will hence turn useful to refresh here the correct test procedure:

1. First, filter with a sieve or cloth the sample under analysis, so as to eliminate fibers and other macroscopic solids.
2. Sample the exact amount of slurry to test with an automatic micropipette, pressing the piston to the first stop, then immerging the tip in the sample to about half its length, and then releasing the piston to aspire the sample.
3. Then load in the test glass 50 g (50 ml) of distilled water (weighted with a scale or measured with a graduated cylinder). Add the 5 ml of filtered slurry with the pipette, this time pushing its piston to the end, so that the sample tip remains completely empty.
4. At this point, launch the automatic titration.

The values obtained with the titrator are three: the FOS parameter (VFA), the TAC parameter (TA), and the ratio FOS/TAC (VFA/TA). All three parameters must be checked as a whole set of information, and not only the nondimensional ratio FOS/TAC. For this scope, Table 1.10 is more useful than Table 1.9.

To get some utility from the FOS/TAC titrator, the test should be performed at least once a day, always at the same time, and the measured values should be monitored together with the energy productivity of the plant, for instance, by means of a spreadsheet and corresponding graphic function. In case some drift of the values becomes evident, (productivity drops, increase of the FOS, and/or progressive sinking of the TAC parameters), then complementary tests will be necessary to detect if there is some problem and

TABLE 1.10

Reference Values of FOS, TAC, and pH in Agricultural Anaerobic Digesters, According to Daniel and Baumgartner

	FOS [mg/l]	TAC [mg/l]	FOS/TAC	pH
Normal operation	2.100	9.000	0.23	7.8
Start of acidification	3.300	8.300	0.40	7.6
Acidified reactor	4.800	6.800	0.71	7.0
Biological collapse	7.300	3.800	1.92	5.6

its causes. The said tests (SMA test; hydrolytic activity test, analysis of the VFA profile, analysis of the oligoelements present in the sludge, analysis of the ammonia and total N, measure of the electric conductivity, etc.) will be explained in detail in Chapters 3 and 4.

Bibliography

Appel, L., Baeyens, J., Degrève, J., and Dewil, R., Principles and potential of the anaerobic digestion of waste-activated sludge, *Progress in Energy and Combustion Science*, 34(6), 755–781, 2008.

Bolzonella, D., *Monitoring and Lab Analysis—Assays*. Lecture of Jyvaskyla Summer School, 2013. http://www.valorgas.soton.ac.uk/Pub_docs/JyU%20SS%202013/VALORGAS_JyU_2013_Lecture%203.pdf.

Garuti, M., Soldano, M., and Fabbri, C., Conducibilità elettrica, utile monitorarla nel digestato, L'Informatore Agrario, 40, 2014.

Gerardi, M.H., *The Microbiology of Anaerobic Digesters*. John Wiley & Sons, Hoboken, NJ, 2003.

Gerber, M., and Span, R., An analysis of available mathematical models for anaerobic digestion of organic substances for production of biogas, *International Gas Union Research Conference*, Paris, France, 2008.

Joan Mata-Alvarez, Biomethanization of the organic fraction of municipal solid wastes, IWA Publishing, London, UK, 2002.

Kosko, B., and Lupoli, A., *Il fuzzy-pensiero: Teoria e applicazioni della logica fuzzy*, Collana Tascabili Baldini and Castoldi, I nani. Vita matematica; trad. di Agostino Lupoli, 4a ed., Baldini & Castoldi, Milan, 2000.

Kuo B. *Automatic Control Systems*. Prentice-Hall International Editions, Upper Saddle River, NJ, 1986.

Lee Sung, J., Relationship between oxidation reduction potential (ORP) and volatile fatty acid (VFA) production in the acid-phase anaerobic digestion process, degree thesis, University of Canterbury, 2008.

Lossie, U., and Pütz, P., Targeted control of biogas plants with the help of FOS/TAC, Application note of Hach-Lange® DOC042.52.20011.Mar08, 2008.

Ogata, K., *Modern Control Engineering*, Prentice-Hall International Editions, Upper Saddle River, NJ, 1985.

Prokopchenko, O.E., The Gompertz Sigmoid function and its derivative, The Wolfram Demonstrations Project, http://demonstrations.wolfram.com/TheGompertzSigmoidFunctionAndItsDerivative/.

Raposo, F., De la Rubia, M.A., Fernández-Cegrí, V., and Borja, R., Anaerobic digestion of solid organic substrates in batch mode: An overview relating to methane yields and experimental procedures, *Renewable and Sustainable Energy Reviews*, 16(1), 861–877, 2011a.

Raposo, F., et al., Biochemical methane potential (BMP) of solid organic substrates: Evaluation of anaerobic biodegradability using data from an international interlaboratory study, *Journal of Chemical Technology and Biotechnology*, 86(8), 1088–1098, 2011b.

Schnürer, A., and Jarvis, A., *Microbiological Handbook for Biogas Plants*, Swedish Waste Management U2009:03, Swedish Gas Centre Report 207, 2010.

Van Lier, J., Sanz, M., and Lettinga, G., Effect of temperature in the anaerobic thermophillic conversion of granular and dispersed sludge, *Water Research*, 30(1), 199–207, 1996.

Varnero Moreno, M.T., *Manual de biogás*, Edition by FAO (Food and Agriculture Organization) and Government of Chile, Santiago, Chile, ISBN 978-95-306892-0, 2011.

Violante, P., *Chimica e fertilità del suolo*, Edagricole, Bologna, Italy 2013.

2

Overview of the Laboratory Methods for the Analysis of Fermentative Processes

---

2.1 Basic Notions of Metrology: Accurateness and Precision or Repeatability

"I often say that when you can measure what you are speaking about, and express it in numbers, you know something about it; but when you cannot measure it, when you cannot express it in numbers, your knowledge is of a meagre and unsatisfactory kind; it may be the beginning of knowledge, but you have scarcely, in your thoughts, advanced to the stage of science, whatever the matter may be."

These words of Sir William Thomson (better known as Lord Kelvin, because of the temperature scale nominated in his honor), published in 1883, summarize the essence of this book. Running a biogas plant requires measuring constantly a series of parameters so as to obtain more methane per cubic meter of digester (optimization of the process) and/or more methane per ton of feedstock (optimization of the substrate). The question is: What does the word "to measure" mean in this context?

> To measure means to compare an unknown magnitude with another magnitude assumed as measurement unit.

This is relatively easy when one intends to measure a well-defined physical unit, for instance, a weight, a length, or an electric current. On the contrary, it is very difficult to measure a complex process, like anaerobic digestion, because we cannot count the amount of living bacteria of each species and their respective metabolic activities. We can only define a *state* of the whole ecosystem composed of the bacteria and the substrate, on the basis of indirect information about their interactions (instant flow and composition of the gas, pH, and so on).

We can add another truth to Lord Kelvin's words, apparently obvious but not always proving true in practice: Apart from *being able to measure*, it is also necessary to *know how to measure* correctly. Hence, before discussing some

Sir William Thomson, baron of Kelvin. (Public domain photo from https://commons.wikime-dia.org/wiki/File:Portrait_of_William_Thomson,_Baron_Kelvin.jpg.)

practical examples, in the next paragraphs, we are going to introduce some theoretical concepts about how to perform reliable measures.

2.1.1 Definitions of Accurateness and of Precision or Repeatability

Figure 2.1 summarizes the definitions of both concepts with the example of a keen sniper who tests four different rifles or, conversely, of four snipers with different degrees of ability testing the same rifle. The snipers can fail hitting the center of the target. The analogy of the snipers within our context allows

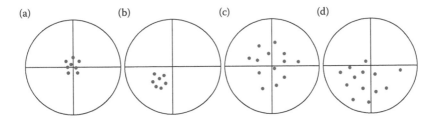

FIGURE 2.1
Graphic analogy of the concepts of accurateness and precision or repeatability. (a) Accurate and precise. (b) Precise, but not accurate. (c) Imprecise (or disperse), but the average value may be acceptable because the dispersion is not biased in a given direction. (d) Imprecise and inaccurate. (Source of the picture and recommended reference: https://en.wikipedia.org/wiki/Accuracy_and_precision.)

us to state that the measurement errors can be caused either by an incorrect calibration of the instrument or by the user's inexperience or carelessness.

A good instrument must be precise, i.e., it must measure values that are consistent in time, and it must be accurate too, i.e., it must measure values very close to the true value of the physical magnitude (case A).

An instrument that is precise but not accurate (case B) induces a systematic error, easy to compensate on the condition that both the amount and the sign of the error are known.

An instrument that is accurate but not precise (case C) induces random errors. Such kinds of instruments are acceptable for maintenance scopes, because the average value of the measures will be close to the true value, but it will require a relatively high number of measures to obtain a representative average value.

An instrument that is neither precise nor accurate (case D) should not be employed, because the measures will be unreliable.

2.1.2 Error Propagation

There are two different situations, in which it is necessary to decide on the basis of measured values. In the first case, we *measure directly* with a *dedicated* instrument the physical magnitude we are interested in knowing; in the second case, the unknown physical magnitude results from a *calculation from one or more measures*. When measuring directly a physical magnitude, the error (or uncertainty) of the measured value depends directly on the instrument (and/or on the user's skill). When the unknown physical magnitude is derived from one or more measures, the total error or uncertainty of the calculated value will be a function (usually a sum or an amplification factor) of the single measures and their respective errors. To understand how the errors propagate when a value is derived from several measures, it is possible to employ a series of rules, which are based on theorems. Since the scope of this book is to prioritize practical experience over theory, the development of the theorems will not be presented here, and the error propagation will be estimated in all cases by the application of the said rules. We are going to start with the definition of the error types and their symbols, presenting a few examples that will help understand how to apply the error propagation rules.

Definition no. 1
$E_{(x)}$ = absolute error or uncertainty of the measured magnitude x.

If x_m is the value of a physical magnitude measured in a given way and x_v is the true value (in general unknown), the absolute error $E_{(x)}$ is defined as:

$$E_{(x)} = x_m - x_v$$

$E_{(x)}$ is hence expressed in the same unit of the measured magnitude (m, kg, V, etc.).

Example of Absolute Error

Consider a graduated cylinder with 100 ml of nominal capacity, graduated with 1 ml divisions. The absolute error of a liquid volume measured with this device is ±1 ml, because the intermediate values cannot be measured with certainty. Although it is possible to observe with the naked eye if the level of the liquid is between two divisions, it is not possible to state if it is in the exact middle.

Definition no. 2
$e_{(x)}$ = relative error or uncertainty of the measured magnitude x.

The relative error $e_{(x)}$ is defined as:

$$e_{(x)} = \frac{E_{(x)}}{x_v}$$

Consequently, $e_{(x)}$ is nondimensional and will be expressed in %.

Example of Relative Error

Consider the same graduated cylinder of the former example. If 20 ml of liquid is measured, the uncertainty of the measure is equal to ±1 ml, the relative error is equal to 1/20, and hence ±5%. If, instead, 100 ml of liquid is measured, the relative error will be 1/100, and hence ±1%.

First Practical Conclusion

When a physical magnitude is measured with whatever instrument, it is good practice to choose the instrument's scale such that it results as close as possible to the estimated value of the magnitude. For instance, if it is necessary to measure volumes in the order of 9–12 ml, it is convenient to get a graduated cylinder with 15 ml nominal capacity, because employing bigger graduated cylinders will induce a bigger relative error.

Now that the concepts of relative and absolute errors of a measured value have been defined, the fundamental theorems of the *Theory of Errors* can be stated as the following practical rules:

1. Relative error of the product of two measured magnitudes, $x \cdot y$

$$e_{(x \cdot y)} = e_{(x)} + e_{(y)}$$

2. Relative error of the quotient of two measured magnitudes, x/y

$$e_{(x/y)} = e_{(x)} + e_{(y)}$$

3. Absolute error of the sum or the difference of two measured magnitudes, $x \pm y$

$$E_{(x \pm y)} = E_{(x)} + E_{(y)}$$

4. Relative error of the average of n measured magnitudes $x_1, x_2, ..., x_n = \bar{x}$

$$e_{(\bar{x})} \approx e_{(x)}$$

The former rule is valid only on condition that the single values $x_1, x_2, ..., x_n$ are of the same order of magnitude as $\bar{x}$, and furthermore that the value of $\bar{x}$ is as close as possible to the estimated value of the magnitude (condition of minimum relative error).

Second Practical Conclusion

From the former rules, it can be deduced that the maximum error made when calculating a value with a formula, based on physical variables measured independently, depends on the formula and on the method employed, hence not only on the precision and accuracy of the instruments.

Conventions for the correct expression of measured values, or of values calculated from measures, and their errors

1. $E(x)$ is always written with a single significant figure.
2. The result of a measure whose $E(x)$ is known is written with all the significant figures up to the same order of the error, rounded.

Let us see some practical examples on how to apply the rules and conventions.

Example 1

Measure the dry matter (DM) of a substance, planning the measurement in such a way that, when employing a simple digital kitchen scale with good criterion, the relative error will be less than 1%

Data:
Accuracy of the scale is ±0.5 g (in general, this value is provided by the manufacturer in the user's manual or in the technical sheet)
   Order of magnitude of the DM (supposed) $\approx 30\%$

By definition:

$$DM\,[\%] = \frac{DW\,[g]}{WW\,[g]}$$

where DW is the dry weight and WW is the wet weight.

$$e_{WW} = \frac{E_{WW}}{WW}$$

$$e_{DW} = \frac{E_{DW}}{DW}$$

Since the same scale will be employed for measuring both the humid and the dry matter (after drying a sample in an oven), the absolute errors of both weights are:

$$E_{DW} = E_{WW} = \pm 0.5\ g$$

Applying the rule of the quotient of two measured values, we obtain:

$$e_{DM} = e_{WW} + e_{DW}$$

Since the goal is to keep the error in the determination of DM smaller than ±1%, the former equation becomes:

$$0.01 = e_{WW} + e_{DW}$$

Since we assumed that:

$$DW \approx 30\%\ WW$$

we can hence write:

$$e_{DW} \approx \frac{E}{0.3 \cdot WW} = \frac{0.5\ g}{0.3 \cdot WW}$$

$$e_{WW} = \frac{E}{WW} = \frac{0.5\ g}{WW}$$

Hence:

$$0.01 = \frac{0.5 \text{ g}}{0.3 \cdot \text{WW}} + \frac{0.5 \text{ g}}{\text{WW}}$$

$$0.01 = \frac{(0.5 + 1.666)\text{g}}{\text{WW}}$$

$$\text{WW} = \frac{(0.5 + 1.666)\text{g}}{0.01} = 216 \text{ g}$$

First Practical Conclusion

It is possible to measure with good accuracy even when employing inaccurate instruments, on condition that the measurement procedure is planned with a correct criterion.

Example 2

A student measures the DW of a digestate sample with a laboratory scale, whose error margin (defined by the manufacturer) is ±5 mg. The wet sample weighs 1 g, and the weight after drying it in the oven is 0.045 g. Estimate the error in the determination of the DM under said conditions.

By definition:

$$e_{DM} = e_{WW} + e_{DW}$$

$$e_{WW} = \frac{E}{\text{WW}} = \frac{0.005 \text{ g}}{1 \text{ g}} = 0.5\%$$

$$e_{DW} = \frac{E}{\text{DW}} = \frac{0.005 \text{ g}}{0.045 \text{ g}} = 11.1\%$$

$e_{DM} = 12\%$ (remember that, conventionally, the error must be rounded to its nearest significant figure).

Second Practical Conclusion

It is of no use having accurate (expensive) instruments if they are employed with a wrong criterion!

Example 3

It is necessary to measure the DM of several biomasses, and then calculate the average DM of the mixture, so as to employ the latter value for defining the diet to feed a digester. The available instrument is a halogen moisture scale that shows the result directly as DM with uncertainty ±1%. The manufacturer declares the said accuracy for samples of, at least, 10 g (but the technical sheet does not state if the said accuracy is constant for the whole range, 0%–100%). According to the manufacturer, the error of the instrument when working in "scale mode" is 0.2% (by an international convention, the said percentage refers to the end of scale or scale limit). The maximum weight reading of the instrument is 100 g.

Calculate the relative error of the total VS in the daily ration when employing the average instead of the single values.

N.B. Please observe that, since the instrument measures directly the DM, expressed in % of the wet weight, its measurement unit is %. Hence, "±1% uncertainty" in this context means that the *absolute error* is equal to "1%."

Values of DW% measured on 10 g samples:

Biomass A = 12%
Biomass B = 38%
Biomass C = 60%

The average is then:

$$\overline{DM} = 36.66\% \approx 37\%$$

Remember: By convention, the calculated value is rounded to the nearest significant figure. If the instrument is unable to detect variations smaller than 1%, i.e., it cannot distinguish between a sample with 36% DW and a sample with 37% DW, it would be absurd to express the measured value with decimals.

Calculation of the Errors

Since the single values do not have the same order of magnitude of their average, the approximate rule derived for the error of the average is not valid in this case. By definition:

$$\overline{DM} = \frac{DM_1 + DM_2 + DM_3}{3}$$

$$e_{\overline{DM}} = \frac{E_{(DM_1+DM_2+DM_3)/3}}{(DM_1 + DM_2 + DM_3)/3}$$

Since 3 is a figure, it has no associated error, so:

$$E_{(DM_1+DM_2+DM_3)/3} = E_{(DM_1+DM_2+DM_3)} = E_{DM_1} + E_{DM_2} + E_{DM_3} = 3 \cdot E_{DM}$$

$$e_{\overline{DM}} = \frac{3 \cdot E_{DM}}{(DM_1 + DM_2 + DM_3)/3} = \frac{0.03}{0.37} = 81\%$$

Please note that performing three separate measures and then calculating their average induce an absolute error bigger than the individual error of each measure, especially if the single values are much different from the average. In this case, the absolute error of the average value is three times the error of the single measure. Alternatively, we can also say that the average value has an uncertainty of 81% (meaning 81%

uncertainty of the calculated 37%). Both forms are equivalent, we must only pay attention not to confound what "percent" means in each case. Summarizing: By individually measuring the DM percentage of each biomass, we calculate an average DM of the mixture as 37%, but in practice the actual mixture could have any percentage of DM in the range of 34%–40%.

Third Practical Conclusion

When calculating average values of measures taken far below the maximum reading of the instrument, and furthermore, if the individual readings are of different order of magnitude than the average, the error tends to be amplified. In our example, it would have been easier, and more accurate, to take 100 g sample from each biomass, mix it very well, then take out at least 10 g of the homogeneous mixture, and measure directly the DM of the mixture. We would have obtained the desired value with just one single test, and furthermore with an absolute error margin of ±1%.

Example 4

Suppose weighing a sample of biomass with a halogen moisture analyzer set in "scale mode." The maximum weighing capacity of the instrument is 110 g. The manufacturer states that the maximum error is ±0.02% and that the resolution is 0.005 g.

The reading of the scale is 22.385 g. Is it correct to adopt this value for further calculations?

Since the scale error is 0.02% of 110 g, then:

$E_{(x)} = 0.0002 \times 110 \text{ g} = 0.022 \text{ g}$

By convention, $E_{(x)}$ must be rounded to its first significant figure; hence:

$E_{(x)} = \pm 0.02 \text{ g}$

The specification of the manufacturer "resolution = 0.005 g," means that the instrument can discriminate between two samples weighing X.XX0 g and X.XX5 g. As we can see, the maximum error of the reading is bigger than the resolution (this a typical feature of low-cost moisture analyzers, because in "moisture test mode" the important feature is the *difference* between the initial and the final weight, independently of the accurateness of the absolute initial and final weights themselves). Practically, in "scale mode" we are not able to know if the real weight of the sample under examination is 22.385 g, because under the measurement conditions of the present example, the actual weight could be any value in the range of 22.365–22.405 g.

The relative error of the reading is then:

$$e_P = \frac{0.02 \text{ g}}{22.385 \text{ g}} = 0.000893 \approx \pm 0.09\%$$

The correct way to express the reading is:

$$x = 22.38 \text{ g} \pm 0.02 \text{ g}$$

Fourth Practical Conclusion

The former example applies especially to digital instruments: The number of decimals shown in the instrument's display does not necessarily imply a higher accurateness of the reading. In the previous example, the displayed reading has three decimals, but only the first two are significant.

2.2 Measurement Methods Employed in the Biogas Industry

We have seen in Chapter 1 that the optimization of a biogas plant requires monitoring a series of physical and chemical magnitudes. Some of them (for instance, the profile of the volatile fatty acids [VFA]) require the services of specialized laboratories, since they count on sophisticated instruments to carry out complicate analytical techniques.

On the other hand, many other important variables can be easily and directly measured *in situ* by the biogas plant manager, with just some standard instruments and without special scientific skills. Said variables provide in real time an immediate idea on the necessary actions to keep the anaerobic digestion (AD) process fully efficient. The practical examples 1 and 2 in the former Section 2.1 are very useful to clear the key concept: If we know *how to* measure with the available instruments, *we can* also measure correctly and so we can *improve* our process. In this section, we will present an overview of the tests that the biogas plant manager can perform, the necessary instruments for each scope, and some guidelines for the selection and purchase of the most suitable instruments. The test methods and the practical application of the results obtained will be discussed in Chapters 3 and 4, respectively.

First, we can classify the *in situ* tests into two big categories: physicochemical tests and biological tests. Physicochemical tests (on biomass, slurry under digestion, or digestate) are somehow preparatory to the biological tests and altogether provide indirect information from which we can easily infer the state of the microbial ecosystem and the quality of the biomass employed as feedstock. Table 2.1 summarizes the more suitable tests to be performed *in situ* by the biogas plant manager, because of their simplicity and the importance of the information obtained with them.

2.2.1 Classical Volumetric and Barometric Methods

The most useful tests for analyzing the AD process are the biological ones. All those shown in Table 2.1 are called *respirometric tests*, because they are based on measuring the quantity of methane (or biogas) produced in a given time under certain conditions, from which it is possible to deduce the necessary information about the living mass and the metabolic activity of the microbial ecosystem.

TABLE 2.1

Useful *In Situ* Tests for Optimizing Biogas Plants

Physicochemical Tests	Measured Variable	Necessary Instrument
DM	% of total solid matter	Halogen moisture analyzer or an oven and a precision scale
Ash (A)	Mineral fraction of the DM (%)	Muffle oven and precision scale
Volatile solids (VS)	% of organic matter in the solids	Calculation: VS = DM-A
COD	% of organic matter (in slurry and liquid substrates)	Spectrophotometer
N ammonia, N total	Dissolved NH_4^+ and total N (sludge and liquid substrates)	Spectrophotometer
pH/ORP	pH/ORP	pH meter and ORP-meter
Composition of biogas	% of CO_2, and by difference, % of CH_4	Syringe, plastic tubes, caustic soda, or portable gas analyzer
Biological Tests	**Measured Variable**	**Necessary Instrument**
BMP	CH_4 productivity of the feedstock / Residual productivity of CH_4 of the digestates	Batch reactors and counters of the accumulated gas volume, ideally also data logger to minimize manual work
AD curve	Kinetics of the AD process (the parameters a, b, and c of Gompertz equation)	Batch reactors and counters of the accumulated gas volume, ideally also data logger to minimize manual work
Specific Methanogenic Activity (SMA)	Capacity and speed of degrading acetate (and/or VFAs in general)	Batch reactors and counters of the accumulated gas volume, ideally also data logger to minimize manual work
Hydrolytic activity	Capacity and speed of degrading starch and cellulose (polysaccharides in general)	Batch reactors and counters of the accumulated gas volume, ideally also data logger to minimize manual work
Proteolytic activity	Capacity and speed of degrading proteins	Batch reactors and counters of the accumulated gas volume, ideally also data logger to minimize manual work
Lipolytic activity	Capacity and speed of degrading lipids (oils and fats)	Batch reactors and counters of the accumulated gas volume, ideally also data logger to minimize manual work

Two "classical" schools of thought exist about how to measure the volume of gas produced per unit of time:

1. *Measure at constant (or almost constant) pressure*: This method is based on the principle that, if the pressure of a gas is kept constant, its volume will be proportional to the mass of gas produced by the AD. Those methods employing the said principle are called *volumetric*, because the measurement device counts the volume of gas produced (e.g., by measuring the displacement of a piston, the displacement of a water column, and any kind of mechanical counter).

2. *Measure at constant volume*: This method is based on the principle that, if the volume of the digester is kept constant, the pressure on the walls produced by the biogas is directly proportional to its mass. Those methods employing the said principle are called *barometric*, since the measurement device (mechanical manometer, mercury column, solid-state pressure sensor, etc.) senses the pressure of the gas produced by the AD.

A third school of thought, limited to a fraction of the academic world, proposes the gas chromatographic method. It is similar to the barometric method, but instead of measuring the increase in pressure, the instrument measures the variations in the composition of the gas. The said method was described by Hansen et al. (2004). It has an acceptable accuracy (about 2% instrumental error), but its sensitivity is low: According to the cited authors, it cannot detect variations smaller than 75 Ncm^3. Furthermore, the gas chromatographer is an expensive instrument requiring a skilled operator and frequent calibration, so it is unsuitable for industrial use.

Now let us analyze some of the possible technical solutions, both hand-crafted and commercial, together with their advantages and disadvantages.

2.2.1.1 Volumetric Methods

Figure 2.2 shows one of the most rudimental systems possible: the displacement of a piston in a conveniently graduated cylinder. It was first adopted by the University of Hohenheim in Germany, and hence it is usually cited in the literature as "Hohenheim system." For its construction, just some syringes are needed, which will be both fermenter and measurement system in one, plus a thermostatic bath to keep the desired temperature (or an electric incubator). The only advantage of this solution is its extremely low cost and that the materials are readily available. Its most important disadvantages are the following:

- The friction coefficient of the piston is not constant, so it will displace jerkily and hence the reading is not reliable.

- The volume of most of the syringes available in the market is usually smaller than 100 ml; hence it is necessary to work with very small quantities of inoculum and substrate. Large extrapolation errors are then induced and in most cases the representativeness of the sample is lost.

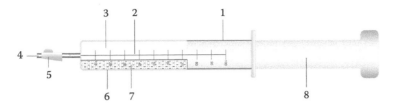

FIGURE 2.2
Volumetric measure test by means of a syringe. (1) Oil or silicone grease (both for lubrication and sealing), (2) Graduations, 1 ml, (3) Volume occupied by the biogas, (4) Output valve for the gas chromatographer, (5) Valve, (6) Glass syringe, (7) Slurry under digestion, and (8) Piston. (Drawing of the author, based on the draft of norm VDI 4630/2014. The set of syringes is placed in a rotating support, inside a thermostatic cabinet or incubator.)

- The small volume of the syringe obliges to take readings quite often. Should the volume of produced gas be bigger than the syringe's own volume, as for instance during the weekend, the piston could be pushed out of the cylinder, losing all the gas and spoiling the test.
- The measurement error is quite difficult to estimate because of the piston's jerky displacement.
- It is mandatory to get a weather station to measure the room's pressure and temperature and be able to normalize each single reading.
- The small diameter of the syringe's aspiration tube limits this method to liquid substrates or very diluted organic matter, for instance, wastewater or nutrient solutions prepared in the laboratory.

Figure 2.3 shows the operation scheme of another easy-to-build instrument: the eudiometer, invented by Alessandro Volta for his experiments with what

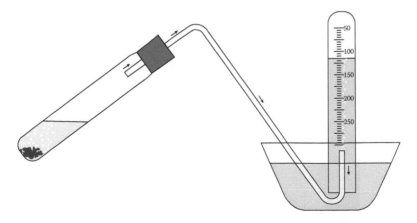

FIGURE 2.3
An elementary eudiometer. (Picture courtesy of http://www.wpclipart.com/science/chemistry/eudiometer.png.html.)

he called "air that burns" (biogas released by decaying organic matter in marshes, a.k.a. marsh gas). In the eudiometer, the "piston" is a liquid column (usually water; oil is seldom employed, although it has some potential advantages over water).

Figure 2.4 shows a homemade eudiometer, and Figure 2.5 shows a commercial bank of eudiometers.

The eudiometer is an instrument invented by Alessandro Volta in 1777 conceived to measure the quantity of gas produced by chemical reactions (quote from Alessandro Volta Foundation, www.alessandrovolta.it). It can be well adapted for performing aerobic respirometric tests. The pros and cons of the eudiometer, when it is employed for anaerobic tests, are summarized in Table 2.2.

Another variant of the volumetric methods is the so-called "fluid displacement". Basically, it consists of two recipients: one working as anaerobic reactor (hence it is possible to add a suitable stirrer and a thermostatic bath for

FIGURE 2.4
"Vintage" eudiometer conserved at CIEMAT of Madrid. (Photo by the author.)

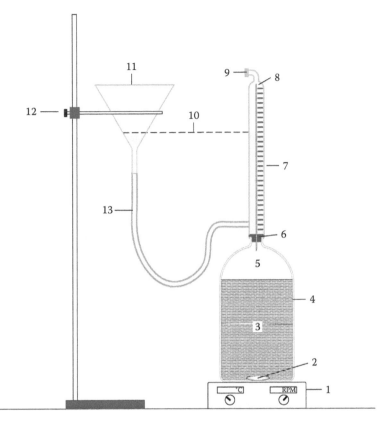

FIGURE 2.5
Eudiometer according to norm DIN 38414. (1) Heated magnetic stirrer, (2) Magnetic bar, (3) Slurry under test, (4) Glass bottle, (5) Headspace, (6) Joint of the eudiometer tube (rubber, plastic, or sanded glass), (7) Graduated eudiometer tube, (8) Gas collection space, (9) Gas sampling and venting port, (10) Water level, (11) Container for collecting the displaced water, (12) Adjustable height stand, and (13) Rubber tube. (Drawing by the author.)

keeping a stable temperature) and another recipient containing liquid, having a siphon through which the liquid displaced by the gas falls in a graduated cylinder. Since the displaced volume is equal to the volume of gas, the accuracy of the measure depends on the accuracy of the graduated cylinder, on condition that, between consecutive readings, the variations of ambient T and P are negligible. According to the norm ISO DIN EN 1042, two qualities of graduated cylinders exist: class A and class B (class B has twice the error of class A). The absolute error depends on the nominal volume of the cylinder and is stated in the label on the device, as can be seen from Figure 2.6. Figure 2.7 shows a volume displacement measurement system built by one of the author's students.

Figure 2.8 shows the scheme of a solution, more sophisticated and expensive than the handcrafted construction in Figure 2.7, based on standard

TABLE 2.2

Advantages and Disadvantages of the Eudiometer for Measuring Biogas from AD

Advantages	Disadvantages
Simple operation, no moving parts	Fragility of the glass tubes.
If the liquid is a solution of NaOH in water, the instrument measures directly the CH_4 produced. If oil is employed, it can measure total biogas with minimum error caused by the solubility of CO_2 in water.	The accuracy of the instrument's reading is proportional to the height of the liquid column, but a high column implies that the operation principle is not any more "at constant pressure," and this induces errors in the normalization of the gas volume.
Parts easily available in the market.	Fabrication requires highly skilled craftsmen; medium to high cost.
Easy reading.	Only manual reading = possibility of human error+much time to dedicate to the test.
Possibility of self-construction.	Impossible to include a stirrer robust enough for agricultural slurry and biomass.
	Requires a special thermostatic bath.
	Readings subject to variations of ambient P and T; so it is necessary to have a thermometer and barometer and also take these readings to compensate the normalization errors.
	Working volume in general is small; hence, during the weekends it could remain dry and spoil the whole test.

laboratory glassware. The accuracy of this second solution will not necessarily be higher than the handcrafted one: Remember that the accuracy is exclusively related to the class of the graduated cylinder measuring the volume of displaced liquid.

Compared to the eudiometer, volume displacement systems present the advantage of easier "DIY construction," (even with common materials, not necessarily laboratory glassware), the possibility of foreseeing a bigger displacement volume (hence no risk of losing the test's result during weekends), constant pressure (because the column height of the NaOH solution will stabilize the pressure), and in general, lower cost. The disadvantages of volume displacement methods are the same already reported for the eudiometer in Table 2.2.

Finally, the family of the volumetric methods includes the volumetric gas counters. In general, these are based on a system of hinged bells immersed in a fluid, or paddle wheels, or small turbines, or even optical systems that count the bubbles released by a small diameter tube submerged in a fluid. Such instruments are more complex, and in general, include some electronic

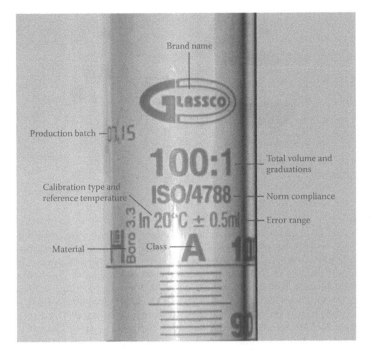

FIGURE 2.6
Example of class A cylinder labeling according to ISO DIN EN 4788. (Photo by the author.)

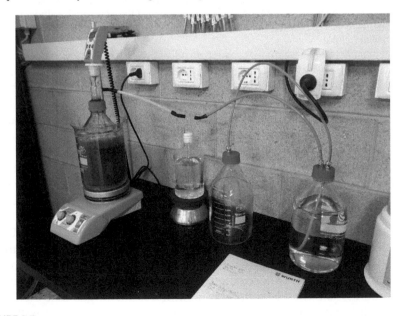

FIGURE 2.7
Volume displacement measurement system built by Mr. Gianluca Bergamaschi, employed to optimize the digester of Azienda Agricola Valsesia s.n.c. (Photo gently provided by its owner.)

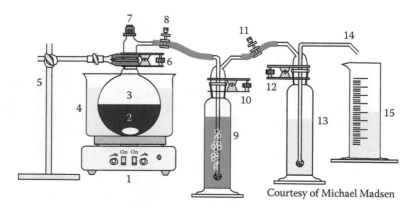

Courtesy of Michael Madsen

FIGURE 2.8
Scheme of a volume displacement measurement system built with standard laboratory glassware, suggested by Dr Michael Madsen. (1) Magnetic stirrer with thermostatic plate, (2) Bottle with magnetic bar for stirring, employed as digester. (The picture shows a laboratory balloon, but in our opinion an Erlenmeyer flask with flat bottom would be more convenient), (3) Digester's headspace, full with biogas, (4) Water bath for maintaining a uniform, constant temperature, (5) Support with clamps, (6) Collar for the flask's adapter, (7) Flask adapter with valve for sampling biogas, (8) Rubber tube with flange or hose clamp, (9) Drechsel bottle (could also be a Muencke bottle or any other similar device for washing gas), filled with NaOH solution and pH indicator (typically phenolphthalein, becoming colorless at pH < 8), (10) Collar for the Drechsel bottle, (11) Rubber tube with hose clamp, (12) Collar, (13) Drechsel bottle filled with colored water to facilitate the readings (the author of Figure 2.8, Dr. M. Madsen, recommends methyl orange as a colorant. Indeed, any other colorant will be suitable, since the scope of coloring the water is just for making it easier to read the graduated cylinder.), and (14) Graduated cylinder. (Picture with Wiki Commons license from http://biogas.wikispaces.com/ Experimental.)

printed circuit and are delivered with a calibration certificate. Their cost is definitely higher than the simple systems we have described so far, but on the other hand their operation is automatic (at least to a certain extent), hence they do not require the permanent attention of an operator. As an example, Figure 2.9 shows the μ-Flow instrument, whose photo is reproduced by courtesy of Bioprocess Control AB.

It is difficult to compile a table with advantages and disadvantages of these types of measurement instruments, since their features vary enormously from a manufacturer to the other. The model shown in Figure 2.9 is the most modern model currently in the market, having been launched in 2013. It includes ambient pressure and temperature sensors, so its reading is already normalized in real time. Other similar instruments do not have this unique feature, so the normalization of their readings must be performed manually. The accuracy varies a lot too, ranging from ±1% to ±5% depending on the brand and model. The resolution (minimum volume of gas that the instrument is able to measure) lies in general in the range from 2 to 20 ml, with some models being unable to detect volumes smaller than 100 ml. The only feature that all these instruments have in common is the manual reading (either

FIGURE 2.9
μ-Flow measure cell. (Photo by courtesy of Bioprocess Control AB.)

on digital display or on an analogic quadrant). Hence, the curves of biogas production obtained with these kinds of counters are usually polygonal. To obtain smoother curves with higher resolution, it is then necessary to take at least four readings per day, and eventually also read the room's P and T (if the counter has no built-in automatic normalization, also a weather station will be necessary). In some models, like the one depicted in Figure 2.9, the counters have standard analogic signal output (4–20 mA or 0–10 V), proportional to the flow of measured gas. In such case, it is then possible to connect the said analogic output to a standard data logger system, so as to continuously monitor the gas production and to obtain data in numeric format (Figure 2.10).

2.2.1.2 Barometric Methods

Most elementary barometric systems consist of a bottle with a cap tightly locked to the bottle's neck by means of a clamp (somehow like a champagne bottle). The cap has a small hole, connecting the gas in the bottle with a manometer by means of a tube passing through. Figure 2.11 shows the concept.

Since the bottle's free volume is constant, because water is incompressible, the ratio between the gas pressure and the volume of biogas produced by the AD, normalized at 0°C and 101.3 kPa, is calculated with the state equation of perfect gases. Under such measurement conditions, this is expressed as:

$$P_r \cdot V_r = m \cdot R \cdot T$$

where
P_r = pressure inside the reactor [Pa]
V_r = volume of gas contained in the reactor [m^3] = $V_{bottle} - V_{sludge}$

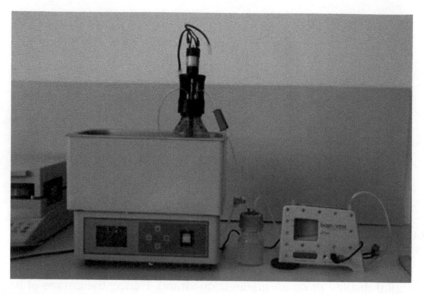

FIGURE 2.10
Low-cost measurement system built by the author with a thermostatic bath, a simple stirred batch reactor, a bottle with NaOH solution, and μ-Flow cell.

m = mass of gas produced by the anaerobic fermentation [mol]
R = universal constant of gases = 8.314 m³ Pa/mol·K
T_r = Absolute temperature of the gas within the reactor [K] (it is assumed to be the same temperature of the incubator)
Since the measured volume of gas must be referred to normal conditions (273.15 K and 101.3 kPa), the former equation becomes:

$$101,300 \, [\text{Pa}] \cdot V_N \left[\text{Nm}^3 \right] = m \, [\text{mol}] \cdot R \cdot 273.15 \, [\text{K}]$$

Our variable is the value V_N.
Under the test conditions, the gas contained in the reactor must satisfy the following thermodynamic state equation:

$$P_r \, [\text{Pa}] \cdot V_r \left[\text{m}^3 \right] = m \, [\text{mol}] \cdot R \cdot T_r \, [\text{K}]$$

The quotient between both members of the equations corresponding to the test and the normalized conditions allows simplifying the product $m \cdot R$. Hence:

$$\frac{P_r \cdot V_r}{101,300 \cdot V_n} = \frac{m \cdot R \cdot T_r}{m \cdot R \cdot 273.15}$$

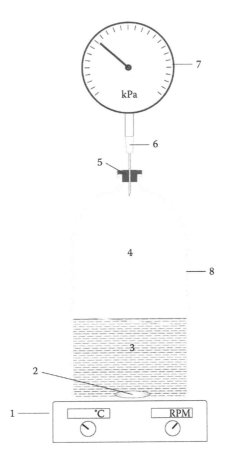

FIGURE 2.11
Elementary barometric system. (1) Heated magnetic stirrer (optional; the bottle can be placed in an incubator and stirred by hand a few times a day), (2) Magnetic bar for stirring, (3) Slurry under fermentation, (4) Collection space of the gas, (5) Hermetic cap, clamped to the bottle's neck, (6) Tube piercing the cap, (7) Manometer, and (8) Glass serum bottle, with mouth suitable for clamping the cap. (Drawing by the author.)

So, the volume of gas produced by the AD, normalized to 0°C and 101.3 kPa, can be calculated with the measured values P_r and T_r and with the known free volume of gas in the reactor, V_r, by means of the following formula:

$$\frac{P_r \cdot V_r \cdot 273.15}{101,300 \cdot T_r} = V_n$$

The formula points out that, since V_n is obtained from a calculation based on three measured quantities, the relative error of the calculated V_n will be equal to the sum of the single relative errors of each measured variable, i.e., of P_r, V_r, and T_r.

Furthermore, the pressure in the headspace is the sum of the partial pressures of all gases contained in it (the N or other gas employed to flush the air, methane, carbon dioxide, and trace gases) and water vapor. The latter is proportional to the temperature, so the water vapor at the test temperature must be discounted before applying the formula for perfect gases (dry gases) already described. If employing a blank to obtain the net methane production by difference, then the systematic error induced by the moisture is null, on condition that the headspace volume is the same both in the blank and in the sample reactors. The following reasoning shows why:

$$G = S - B$$

where
    G = net gas production of the substrate (expressed as pressure increment)
    S = gross gas production of the sample reactor
    B = gross gas production of the blank reactor
    Observe that, according to Dalton's Law of perfect gases, both S and B are the sum of the partial pressures of gas in the headspace volume, S_g and B_g, and water vapor, S_v and B_v; hence:

$$G = S_g + S_v - (B_g + B_v)$$

If the headspace is the same in both reactors, then $S_v = B_v$, so the systematic errors cancel each other and G is really the net biogas production.

We can extend the same reasoning to volumetric methods, because Dalton's Law can be applied either to partial pressures at constant volumes or partial volumes at constant pressure.

Practical Conclusion
Having the same headspace in both the sample and blank reactors is a necessary condition for minimizing the error of biological tests. The sensitivity to the error induced by the moisture content in the gas is different for barometric and volumetric methods: in the first ones, the error induced by eventual differences of the headspace is big, because the headspace is big. In volumetric systems, the headspace is small, and the partial volume of water vapor (at 38°C) is less than 6% of it, resulting in a systematic error that is below the detection limit of most volumetric measurement devices.

As it was the case for volumetric methods, apparatuses based on the barometric method can be either self-constructed, (which means procuring adequate sensors, assembling the reactors, and calibrating each of the latter) or purchased in the market as pre-assembled kits. In general, these latter ones are devices without stirrer, in which the pressure variations are measured with electronic sensors and processed by a software. Figure 2.12 shows an example of self-built reactors with mechanical manometers, and Figure 2.13

shows as an example a barometric system, usually employed for biological oxygen demand (BOD) measures and occasionally also for biochemical methane potential (BMP), produced by VELP.

The disadvantages of barometric methods compared to volumetric ones can be summarized as follows:

1. Because of their own nature, barometric methods only measure the total biogas produced. Hence, it is necessary to perform an additional measure to determine the concentration of methane. This adds error (2% to 5%) to the final value; consequently, they are less accurate compared to volumetric methods measuring directly the net methane production through a CO_2 trap.

2. Barometric systems never have stirrers (in the best case they have just a magnetic bar stirrer). Stirring is a very important factor in AD; hence this fact causes discrepancies between the production measured in the laboratory and the effective yield of the plant (see Chapter 6). Please note that biogas plants work at (almost) constant pressure, i.e., in conditions that are more similar to those of volumetric methods.

FIGURE 2.12
Example of homemade barometric reactors employed by the company Biobooster (Italy) during their start-up research. (Photo by the author.)

FIGURE 2.13
Barometric bottles, a device originally conceived for *aerobic* respirometric measures. Such systems are usually placed in a thermostatic cabinet and have no stirring. The communication between the pressure sensor and the datalogging unit is wireless. (Photo by the Author, with thanks to Dr. Diana Castro Anca, SOLOGAS, Spain.)

3. CO_2 and CH_4 have different solubilities in water. When the pressure in the reactor rises, part of the produced CO_2 dissolves in the aqueous phase. This distorts the measured values in two ways: On one side, the proportion of carbonates in the slurry changes, influencing the alkalinity and, ultimately, the kinetics of the organic matter's degradation. On the other side, as CO_2 dissolves in the slurry, the concentration of CH_4 in the collected gas rises, giving the false impression that the biogas quality obtained is higher than the actual one.

4. Barometric methods are more suitable for *aerobic tests*, for instance, the determination of the BOD of sewage water. They were indeed born for the said scope and readapted a second time to measure BMP.

5. Sometimes the samples produce more biogas than foreseen; hence the reactor's pressure rises too much and, if for any reason the safety valve does not release the excess gas, we risk finding our laboratory as shown in Figure 2.14.

6. Some barometric systems automatically release overpressure and the total biogas is then computed as the sum of n independent measures of pressure and gas composition. Since each measure has its own error, the total final absolute error is n times higher than the individual absolute error.

FIGURE 2.14
Unpleasant consequences of a reactor's explosion caused by overpressure. (Photo by Åssa Jarvis, 2010.)

7. Barometric systems sending the pressure signal to the datalogger via wireless interface sometimes lose the connection because the battery in the sensor unit gets exhausted before the end of the test. Losing data during a test makes the results unreliable.

2.2.2 Automatic Measurement Systems: AMPTS and BRS

The techniques for measuring the volume of biogas produced by AD remained almost unchanged from the first experiences conducted by Alessandro Volta with the "air that burns," and from Pasteur's respirometers for studying the fermentation of glucose. In the last decade, the growth of the biogas industry has increased the need for the maximum standardization of the tests and the simplification of the procedures, so as to minimize manual work and human errors. The most modern systems currently available in the market (2017), unlike their predecessors, were specifically conceived to measure anaerobic fermentative processes and hence largely employ the digital technology and contain a series of innovations that turn them particularly useful for the biogas plant manager. The said instruments are the AMPTS II (and its smaller version, called AMPTS Light) and the BRS. AMPTS means *Automatic Methane Potential Test System*, and BRS means *Bio Reactor Simulator*. Both the reactors and the measurement methods of said instruments have been standardized to the maximum extent, so as to provide highly repeatable results without the need of engaging specially skilled personnel for the laboratory (Figures 2.15, 2.16, and 2.17).

The said instruments share different functional features and even their external appearance, but their scopes in the analysis of fermentative processes are quite different. Their common features are:

1. Measurement method: liquid displacement counter.
2. Real-time normalization of the measured volume: all of them include sensors of P and T in their electronic circuits and a normalization software that considers even the relative moisture of the gas in the reactor's headspace and the composition of the flush gas.
3. Highly automatized calibration protocol (maximum error ±1%, Measurement resolution ±10 ml).
4. No need of periodical recalibrations.
5. Remote access to the data through LAN or Internet. No need to go physically to the laboratory for checking your experiment.
6. Interactive software based on the Hyper Text Markup Language (HTML) format. This means that the interface for accessing the instrument can be any Internet browser. Such a feature has the following advantages: no need to have a dedicated PC, since it can be accessed even through a smartphone, nor is it necessary to install any special software, since its operation is independent of the operative system.

FIGURE 2.15
The AMPTS II. (Photo by courtesy of Bioprocess Control AB.)

FIGURE 2.16
The AMPTS Light, a smaller and cheaper version of the AMPTS II, which is more suitable for industrial use. (Photo by courtesy of Bioprocess Control AB.)

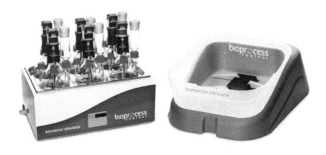

FIGURE 2.17
The BRS. (Photo by courtesy of Bioprocess Control AB.)

7. Data output both graphically (in the PC's screen) and in numerical format (eXtensible Markup Language [XML] and comma separated values [CSV] tables, easily readable with any spreadsheet).

From the point of view of the biogas plant's manager, the most important aspect to know about these instruments is their practical application. The AMPTS II has 15 channels measuring the gas flow from as many 500 ml batch reactors. It is an instrument conceived for measuring the BMP of biomasses (for instance, with the scope of determining their market price based on their effective energy potential). So, the AMPTS II is built and calibrated to measure *directly* the net amount of methane produced, by cleaning the biogas through filters of caustic soda that eliminate the carbon dioxide. As shown in the beginning of this chapter, measuring *directly* a given magnitude provides higher accuracy of the results than measuring several variables and then calculating the result with a formula. The AMPTS Light is analogous to the AMPTS II. The differences are just that it has only 6 channels and 21 batch reactors (more robust and requiring less accuracy in the preparation of the feedstock sample, so as to facilitate its use in the industrial environment). The BRS is an instrument more suitable for scientific research (universities, constructors of biogas plants, consultancy companies). It is an instrument that, unlike the former ones, requires a permanent connection to the *Cloud* and hence has an infinite virtual memory. Consequently, it is suitable for performing long-term continuous tests, and its software is prepared to store and process the relevant parameters for the said scope: the organic loading rate (OLR) and the instant gas flow. The BRS is particularly suitable to analyze dynamic operation conditions:

1. To simulate transient states (e.g., the possible strategies for starting a new digester or the restart after a biological collapse);
2. To analyze possible modifications to the operation of the plant (e.g., to check if, after changing the digester's diet, it may be convenient or not to switch from mesophilic to thermophilic operation, and how to carry out the said change without altering the productivity of the plant);
3. To simulate the operation of a multistage plant, or the influence of a different OLR...

Summarizing: The AMPTS Light is suitable for managing one or two biogas plants and for determining the fair price of biomasses. The AMPTS II is suitable for research purposes, for certificating the quality and methane yield of waste organic matter and commercial feedstock, and for the centralized management of several plants (it is possible to manage easily two to five plants, assuming these are within 100 km from the laboratory. Bigger distances make the transport of the samples more expensive and complicated, especially if these are quickly putrescible matters). The BRS operation is outside of the scopes of this book, being mainly conceived for scientific and industrial research (universities, constructors of biogas plants, specialized consultancy, and engineering companies and owners of very complex biogas plants working with variable loads).

2.2.3 Reactors for Biological Tests: Which Are Better?—Big or Small Ones?

The argument of the reactor's dimensions is always an occasion for debate, sometimes even "intense." Some manufacturers and exponents from the academic world will assert that "the bigger, the better." Some individuals will affirm that "20 and 50 l is the standard in Germany" (a false argument, because the German norm VDI 4630 provides a guideline to choose the size of the reactor, depending on the test's scope, but does not state a concrete volume; see Chapter 6). We will see in Chapter 3 that the error in the determination of a feedstock's BMP is independent from the volume of the reactor (when employing volumetric methods), and in Section 2.2.4 we will introduce the errors generated by the test conditions, which are independent of the reactor's volume too (if volumetric methods are adopted). The belief that a big reactor volume increases the accuracy of the tests lies in the old technologies for measuring gas flows with instruments similar to that of domestic natural gas counters, characterized by fair accuracy in a given flow range, but lacking sensitivity in the case of extra low flows. Modern laboratory flow meters available in the market have a high sensitivity, in the range of 2–10 Ncm^3/h. Hence, it is useless to employ reactors bigger than 2 l (batch) or 5 l (continuous), because the gas flow from the most frequent substrates will be high enough to avoid detection problems. Big volume reactors are required for some very specific cases, like the research on the AD of very refractory substrates (for instance, pesticides and biodegradable plastics) and low concentration wastewater, or to the case of extremely heterogeneous substrates, like organic urban waste. In the first case, it is necessary to work with a big reactor in order to produce a quantity of gas big enough to be measured in reasonably short times, unless an instrument with high sensitivity is employed to measure the gas flow. In the second case, a big reactor turns out to be more suitable for digesting a volume of sample big enough that can be considered "representative" of a full truck of very heterogeneous garbage. The latter problem could anyway be solved with just a 10 L reactor and some statistic processing of the data.

In the special case of managing an agricultural or industrial biogas plant, reactors from 500 ml to 2 l are more than enough to cover all possible routine tests.

FIGURE 2.18
15 l reactors bank in use at CIEMAT, Madrid. (Photo by the author.)

Figure 2.18 shows an old bank of reactors, 15 l each, belonging to CIEMAT, Madrid, nowadays employed only for some special research on scarcely degradable substrates.

2.2.4 The Most Frequent Errors in the Measure of Small Gas Flows

2.2.4.1 Normalization

If an instrument with built-in normalization in real time is not available, measuring correctly requires reading the volume of gas (or the internal pressure of the reactor, if a barometric method is employed) and also of the ambient P and T in that moment, so as to apply the state equation of perfect gases and obtain the normalized value accordingly. This method implies the addition of the relative errors of the gas volume (or pressure) measurement system, of the ambient temperature (or the temperature of the incubation chamber of the reactor) and the atmospheric pressure. It is common practice (especially in older publications) not to perform said corrections when taking each individual reading, and to assume a standard pressure equal to 101 kPa and 20°C as constant laboratory conditions for all readings. This introduces an additional error in the order of 6%–12%, since the atmospheric pressure can vary more than 60 mbar in 1 month and the temperature, even in a research laboratory, varies more than 6°C, especially during weekends (Strömberg et al., 2012).

As explained in Section 2.2.1.2, the formula for the normalization of volume in the case of measures taken with barometric methods is:

$$\frac{P_r \cdot V_r \cdot 273.15}{101.000 \cdot T_r} = V_N$$

With the same reasoning, it is possible to deduce that, in the case of volumetric methods, the normalization formula becomes:

$$\frac{P_{atm} \cdot V_{mis} \cdot 273.15}{101,000 \cdot T_{amb}} = V_N$$

where

P_{atm} = atmospheric pressure (measured with an ambient barometer) at the moment of reading the volume of gas (Pa)

V_{mis} = measured volume of gas (m³)

T_{amb} = ambient temperature in K (273, 15 + T in °C)

On applying the rules of error propagation, even in the case that each single reading is normalized on the ambient conditions of the moment, the normalization of the gas volume introduces a relative error equal to:

$$e_{V_N} = e_{barometer} + e_{thermometer} + e_{gas\ counter}$$

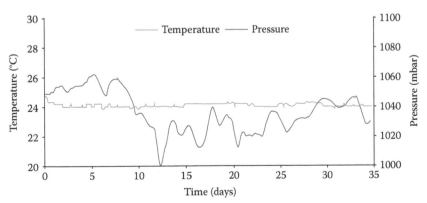

Variations of ambient P and T during a typical month in the laboratory of the University of Lund. (Courtesy of Prof. Jing Liu.)

If the experimental set does not have a CO_2 trap, then an additional measure is necessary—the composition of the biogas. Hence, the total relative error becomes:

$$e_{V_N} = e_{\text{barometer}} + e_{\text{thermometer}} + e_{\text{gas counter}} + e_{\text{gas analyzer}}$$

2.2.4.2 Correction of the Moisture

The headspace of a digester is filled with saturated biogas, i.e., having 100% relative moisture. Under the usual temperature conditions of biogas plants (in the range of 36°C–50°C), the total volume of gas contains 6%–10% of water vapor. In general, the instruments and the simplified formulas presented in the former paragraphs do not take into account the effect of moisture, inducing then further error. Remember that the state equation of perfect gases described in the former paragraphs, although being quite accurate in the range of pressures and temperatures near ambient, is only valid for *dry gases*. As explained in Section 2.2.1.2, moisture induces a systematic error that is automatically cancelled on the condition that the headspace in both the blank and the sample reactors is the same.

2.2.4.3 Elimination of Carbon Dioxide

When employing instruments without caustic soda filters (typically barometric instruments, but also some volumetric ones), the measured value represents the *gross* amount of biogas produced, having extremely variable methane and carbon dioxide proportions (the assumption of 60% CH_4 and 40% CO_2 is purely conventional). To obtain a correct value of the methane concentration, it is then necessary to analyze the gas by means of a gas chromatographer or any other type of gas analyzer. The first one provides high accurateness, but results in being too expensive for a field laboratory in the biogas plant. A cheaper alternative is an instrument with electrochemical cells (affordable cost, easy operation, but requiring frequent recalibrations), or an instrument with solid-state sensors (intermediate cost, but requiring periodical substitution of the sensors every 2 or 3 years and recalibration). Determining *a posteriori*, the methane concentration induces an additional error, since it requires two independent measures—the quantity of biogas produced and its composition. The application of a coefficient "from literature" (e.g., assuming 55% of methane if the digested biomass is composed of polysaccharides, and 60% if it is composed by proteins) induces an error too coarse to be acceptable. It is more convenient to plan the experiments in such a way that a caustic soda filter ensures measuring the *net amount* of methane produced by the reactor. It is furthermore advisable to throw away the caustic soda solution at the end of each assay, and to start the next with fresh solution, so as to avoid the crystallization of sodium carbonate (Co_3Na_2) in the filter. Another cheap method to determine the composition of the biogas is the "syringe method" described in Chapter 5, Section 5.2. Its accuracy is limited by the accuracy of the syringe.

2.2.5 Auxiliary Physicochemical Tests

2.2.5.1 *Measure of the pH and the ORP*

Two ways of measuring the pH exist—by means of paper strips impregnated with substances that change color depending on the pH, or by means of a pH meter. The first method is not advisable in biogas plants, because both sludge and digestate tend to color the paper in brown or black, making the reading impossible. For measuring *in situ*, it is advisable to employ a digital pH meter, even a cheap pocket model, like the one shown in Figure 2.19, that measures pH, temperature, and electrical conductivity.

We have seen in Chapter 1 that the pH value is only a coarse parameter, because it does not provide much information about the actual state of the

FIGURE 2.19
Example of pocket combo meter (pH, temperature, electric conductivity, total dissolved solids). (Photo by the author.)

AD process. It is anyway interesting to monitor it; hence the pH meter must be kept perfectly efficient—it is mandatory to purchase the cleaning and calibration solutions for the electrodes, calibrate periodically, and follow the manufacturer's instructions. The instrument must be thoroughly washed after each use, so as to avoid the formation of dirt deposits on the electrode.

All those prescriptions already said about pH meters apply to ORP meters too.

2.2.5.2 *Determination of the* DM and VS *of Biomass*

The determination of the DM of any stuff is relatively easy to carry out. All the instruments required are a scale—if possible with 1 mg error margin or less—and an electric oven (even a small "convection oven" is enough, because the temperature must not exceed 110°C). A halogen moisture analyzer (a special type of laboratory scale consisting on a system for drying the sample by means of a halogen lamp and the necessary electronic circuitry to automatically sense the weight variation and determine the end of the test; Figure 2.20) is easier to use, since it performs the test automatically. In the first case, the duration of the drying phase will be, in general, long. The easiest way is to place the samples in the oven during the evening and leave them overnight. In the second case, the test can last between 15 and 60 min, depending on the moisture content of the sample.

If no oven and no scale are already available, it is convenient to purchase a moisture analyzer with an acceptable error margin (±5 mg or less) so as to employ it for the ash determination too. A moisture analyzer occupies less place and consumes less energy than an oven, allows for quicker and automated tests (less probability of human errors), and can also be used as a laboratory scale. An infinity of models are available in the market, having different functions, accuracy, sensitivity, and of course, prices.

The determination of the VS requires three weightings: the wet sample, the dry sample, and the ash that remains after the calcination. A muffle oven is required to calcine the biomass samples, i.e., to reduce them to ashes. The standard procedure is to heat the muffle to 550°C, i.e., the self-ignition temperature of any organic matter. The sample of dry biomass can be introduced in the oven once its temperature is stabilized. All carbon, hydrogen, oxygen, and nitrogen compounds will volatilize, while the ash will remain in the plate or crucible. The direct combustion of the sample on a flame is not acceptable for several reasons, mainly because a flame's temperature is almost double the self-ignition temperature of biomass, and will partly volatilize the ash, giving a wrong result. Figure 2.21 shows a model of an electric muffle oven. If considering to purchase a muffle oven, it is necessary to remember that two different technologies exist—those with heavy refractory inner lining and those built in light alumina silicate, having just the floor of the inner chamber in hard ceramic material. In the first case, the muffle will have higher thermal inertia (hence higher temperature stability) and its inside will be more resistant to abrasion, small shocks, etc., but will consume more energy to reach the required temperature. The disadvantage of this technology is that its startup requires a series of heating and cooling cycles having growing final temperatures, with the scope of dehydrating the refractory lining and mortar (otherwise the vapor pressure could burst and damage them). The ovens built with inner lining in alumina silicate are lighter, do not require a start-up procedure, but the material is fragile and soft, so it must be treated delicately—avoid scratching and shocks. Since the second type of ovens is produced from a single insulant block characterized by low thermal mass, they heat up quickly and in general consume less energy than refractory ovens. In any case, before buying any given model, it is necessary to check that it has a fume escape duct or hole, because this feature is not always standard.

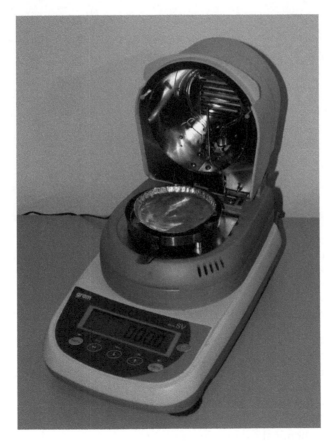

FIGURE 2.20
An example of halogen moisture analyzer, for the automatic determination of DM and moisture, and for laboratory precision weighing in general. (Photo by the author.)

FIGURE 2.21
Example of a muffle oven. The small ceramic tube on the top allows the evacuation of fumes. Air enters through some holes placed under the door's insulation. (Photo by the author.)

2.2.5.3 Determination of the COD of Sludge and Liquid Substrates

Sludge (in general, the content of a digester, the digestates, liquid manure, industrial slurry, and similar substances) usually contains less than 10% of solids. Fibreless slurry (for instance, sludge from wastewater treatment plants, whey, blood...) usually contains too little solids for the easy determination of the organic matter with the method of the VS, because of the large quantity of water to evaporate. In such cases, it is more convenient to measure the chemical oxygen demand (COD). This is a highly standardized chemical test, and pre-dosed kits exist in the market to perform it with the help of automatic instruments, allowing even unskilled operators to perform accurate and reliable tests. The most widely adopted method is known as spectrophotometry. Its working principle is based on the experimental fact that the absorption of light is a phenomenon of interaction between electro-magnetic waves and matter. When a light beam travels across a substance, part of the electromagnetic radiation can be absorbed by atoms, molecules, or crystalline lattices.

If pure absorption takes place (neither reflection nor refraction), the absorbed fraction of light depends both on the length of the optical path—across matter—and on the physicochemical properties of the substance, according to the Lambert–Beer law:

$$A = -\log\left(\frac{I}{I_o}\right) = \varepsilon_\lambda \cdot c \cdot d$$

where
  A = absorbance (a dimensionless coefficient depending on the substance)
  I = Intensity of light at the exit of the optic path
  I_o = Intensity of light at the entry to the optic path
  $\varepsilon\lambda$ = molar extinction coefficient at the wavelength λ
  c = molar concentration of the substance
  d = optic path across the substance
  The working principle of a spectrophotometer is sketched in Figure 2.22.

The liquid under test is dosed in the vial, reacts chemically with the content of the latter, changing the color of the solution. After thermal reaction at high temperature, the vial is introduced in the spectropho-tometer, between a light source of given wavelength and intensity and

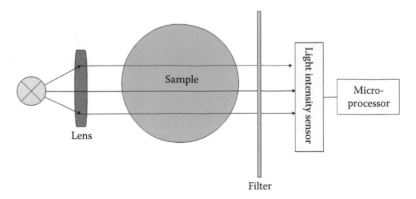

FIGURE 2.22
Sketch of the spectrophotometer's working principle.

the light sensor. The difference between the light intensity across the vial and a reference vial called blank (loaded with distilled water) is then digitally processed by a microprocessor, and the reading appears in the instrument's display.

Figure 2.23 shows a model of spectrophotometer whose vials come from the factory with an identification barcode. In this way the operator needs only to introduce in the vial the prescribed amount of sample (using for said operation an automatic pipette that samples exactly the necessary amount), then shake it gently, and place the vial in the instrument (or in the electric heater and then in the instrument, depending on the test). The spectrophotometer will then read the barcode, ask for confirmation, and proceed to measure the COD (or the parameter identified by the barcode). This highly automatized procedure eliminates the possibility of human errors of traditional spectrophotometers, as for instance, taking a vial for the measure of total nitrogen and reading the measured value as COD.

Spectrophotometers can be either multiparameter or monoparameter. The most interesting parameters for the biogas plant manager are COD, total nitrogen and ammonia nitrogen, and if possible, also total phosphorous. The possibility to also measure total nitrates can be interesting for biogas plants employing the digestate as fertilizer.

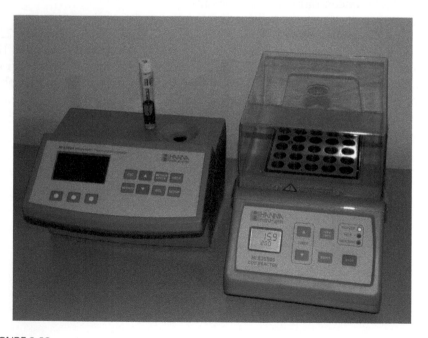

FIGURE 2.23
Multiparameter spectrophotometer for wastewater, cuvette with barcode for automatic measure range setting, and electric furnace for pretreating the cuvettes—a pretreatment required by some tests like COD and Ammonia Nitrogen. (Photo by the author.)

2.2.6 The *In Situ* Laboratory for the Real-Time Control of the Biological Process

From a practical point of view, the minimum laboratory requirements for monitoring the biological processes in the biogas plant should be the following instruments:

1. Automatic moisture analyzer scale (at least ±5 mg resolution)
2. Muffle oven (at least 2, 5 l capacity, temperature control at least till 650°C, accuracy ± 10°C)
3. Combo pH meter, pocket type (pH/temperature/electric conductivity)
4. Oxidation reduction potential (ORP) meter (optional)
5. Accessories: plates, crucibles, tongs, high-temperature protection gloves, scissors, brushes for cleaning bottles and flasks, several funnels of different sizes, stainless steel sieve, safety eyeglasses, single-use gloves, safety eye washing kit.
6. At least three kits for measuring the volume of gas produced by the AD (one blank and two replicates), including reactors with stirrer, thermostatic bath, caustic soda filter, and a gas flow measurement device (preferably with built-in gas volume normalization function)
7. Graduated cylinders (preferably class A) of 50, 100, and 250 ml
8. Spectrophotometer (at least for COD), automatic sampling pipette of the volume recommended by the instrument's manufacturer (usually 200 µl), kit of pre-dosed vials, table heater for the conditioning of the vials (this one is optional; the muffle oven could be used instead, but its temperature control is less accurate and consumes more energy).
9. Working plane: at least two tables lined with Formica or melamine sheet, 60 cm × 120 cm. The surface must be rigid and stable.
10. A washbasin (stainless steel or porcelain), preferably equipped with a dripper, connected to a water line and drain sink.
11. Consumables: single-use gloves, soap and detergents, caustic soda (NaOH), distilled water, plastic bottles for slurry samples, reference substrates for biological tests (acetic acid, starch, gelatin, cellulose, optionally also propionic and butyric acids).

The room where to install the laboratory can be any small room (12–15 m²), even a box inside an industrial building, built with plasterboard or any other kind of dry panels, or a prefabricated box, or a 20' container with insulated walls (like those employed as temporary offices in construction yards). It is important that the door has a key or lock, to avoid non-authorized access, and that the working space is well ventilated, by means of windows or an electric aspirator (especially during the calcination of biomass samples with

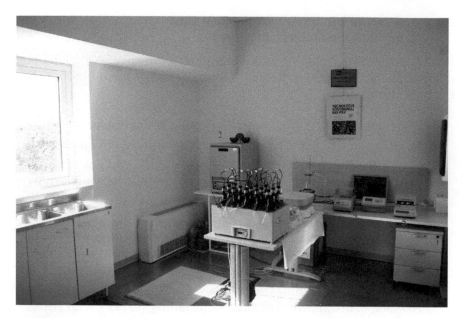

FIGURE 2.24
Model laboratory for BMP analysis in the biogas plant. (1) Washbasin with dripper, (2) Muffle oven, (3) AMPTS II, (4) Spectrophotometer and electric heater for the vials, and (5) Moisture analyzer scale.

the muffle oven). The total electric power available must be at least 3 kW (typically the muffle oven will require about 2 kW). Figure 2.24 shows a model laboratory equipped according to the former indications, installed in a room, about 12 m².

Bibliography

Arrojo, Días, Dampé, Ixtaina, *Medidas eléctricas*, notes of the course on electric measures, Universidad Nacional de La Plata, Buenos Aires, Argentina, 2013.
BIPM (Bureau International des Poids et Mesures), Evaluation of measurement data. Guide to the Expression of Uncertainty in Measurement, JCGM 100, First edition, 2008.
Bolzonella, D., *Monitoring and Lab Analysis—Assays*. Lecture of Jyvaskyla Summer School, Sweden, 2013.
Esteves, S., Devlin, D., Dinsdale, R., and Guwy, A., *Performance of various methodologies for assessing batch anaerobic biodegradability*, International IWA Symposium on anaerobic digestion of solid Wastes and Energy Crops, Vienna, Austria 2011.

Ficara, E., Scaglione, D., and Caffaz, S., Test di laboratorio per valutare la producibil-ità di biogas, Chapter 9. *Biogas da agrozootecnica e agroindustria*, Dario Flaccovio Editore, Palermo, Italy, 2011.

Hansen, T.L., Schmidta, J.E., Angelidaki, I., Marca, E., La Cour Jansen, J., Mosbæka, H., and Christensen, T.H., Method for determination of methane potentials of solid organic waste, *Waste Management*, 24, 393–400, 2004.

ISO Norm 5725/1994 corr. 1998. Accuracy (trueness and precision) of measurement methods and results-Part 1: General principles and definitions.

Nistor, M., Strömberg, S., and Liu, J., Automatic methane potential test system II (AMPTS II) as a tool for screening and selecting organic substrates for anaero-bic digestion processes, Proceedings of the Fourth International Symposium on Energy from Biomass and Waste, San Servolo, Venice, Italy, 2012.

Schnurer, A., and Jarvis, A., *Microbiological Handbook for Biogas Plants*, Swedish Waste Management U2009:03, Swedish Gas Centre Report 207, 2010.

Strömberg, S., Nistor, M., and Liu, J., Standardization as a key to reliable gas mea-surements in biochemical methane potential (BMP) tests, *Proceedings of the Fourth International Symposium on Energy from Biomass and Waste, San Servolo, Venice, Italy*, 2012.

Thomson, W. (Lord Kelvin); *Lecture on "Electrical Units of Measurement"*, Popular Lectures Vol. I, p. 73, May 3, 1883.

VDI (Verein Deutscher Ingenieure), Norm VDI 4630/2006-Fermentation of organic materials. Characterisation of the substrate, sampling, collection of material data, fermentation tests. Beuth Verlag GmbH, Berlin, Germany, 2006.

3

How to Perform Tests under Optimum Conditions

3.1 Scope

The former chapter explained the most suitable instruments and tests used for the management of biogas plants, i.e., those techniques that are useful for optimizing the whole anaerobic digestion (AD) process. This chapter is a practical guide for carrying out tests in the biogas plant, enriched with some tricks and hints to avoid making the typical errors that all beginners do. The practical application of the results obtained with such tests will be explained in detail in Chapter 4.

3.2 Measuring the DM and VS

Two tests are fundamental in optimizing the AD process: the determination of the DM and the volatile solids (VS) of the plant's feedstock (basic information to define the strategies for optimizing the substrate) and the measure of both the VS of the mixture fed to the digester and the digestate exiting it (useful information for optimizing the process).

The determination of both DM and VS are extremely simple tests to carry out, but their importance is fundamental to understand the feedstock's quality and the efficiency of the AD process. It is evident that the feedstock's quality depends on its DM content, because moisture does not contribute to the production of biogas but increases the transport costs. A high moisture content sometimes favors the growth of molds, which generate mycotoxins capable of inhibiting the AD process. Indeed, the main ingredient for the production of penicillin another antibiotics are some selected mold species. In countries where AD feedstock is a commodity derived from energy crops, like Italy and Germany, it is a common market practice to define the price on just the DM content. Such criterion is reductive because two lots of the same feedstock may have the same DM content, but different methane potential. The reasons can be either a different degree of lignification of the biomass, or a higher ash content, or both. In practice, measuring the ash content is useful because the difference between DM and ash, called VS represents the *potentially degradable* fraction of the organic matter, and based on the said datum it is possible to have a coarse estimation of its biochemical methane potential (BMP). The words "potentially degradable" are highlighted because, even if having the same VS content, two lots of the same substrate can have different BMP, depending on the lignin content, on the presence of eventual inhibitors (molds, pesticides, etc.), on the C/N ratio, on the particle size, or depending on pretreatments (thermal, chemical, or biological) that the substrate may have undergone. For all the said reasons, the following is the necessary test sequence for a correct analysis of the substrate's quality (methane potential and degradation time):

determination of the DM → determination of the ash content (and hence the VS) → determination of the BMP

The second application of the VS determination is the measure of the digestion efficiency. Since the quantity of biogas produced by the AD is directly proportional to the VS content, the ratio between the VS of the feedstock and the VS of the digestate is hence an index of the conversion efficiency of the biomass into methane (and hence an indicator of the good or bad operation of the plant).

The sequence of the necessary tests to analyze the plant's overall efficiency is the following:

determination of both feedstock and digestate's VS → calculation of the digestion efficiency → determination of the digestate's residual BMP (complementary test)

3.2.1 Measuring the DM with a Moisture Analyzer

This is the simplest test on any feedstock and requires just following the user's manual of the instrument. The following is the sequence:

1. Place a dish on the scale and bring the reading to zero.
2. Load uniformly the dish with the quantity recommended by the instrument's manufacturer (usually 10 g).
3. Close the heating lid and press the start button.
4. The instrument will start the drying cycle and stop it automatically when it does not detect any weight variation within a preestablished time. On reaching the said point, it will store in its memory the initial and final weights and the resulting moisture content.

Although the moisture analyzer is a very simple instrument, it can easily induce the beginner in making errors, wasting time, and hampering the quality of the results. The following list shows the most frequent errors when measuring DM and VS.

3.2.1.1 Confusing the "Moisture" and the "DM" Scales

By default, moisture analyzers measure moisture (i.e., the weight difference between wet and dry sample) in percentage. Most commercial models allow selecting different ways of expressing the result: Moisture in % (of the wet weight; WW), DM in % (of the WW), ATRO moisture and ATRO dry matter. The word ATRO comes from the German *Absolut TROcken* (absolute DM), and is a conventional way of measuring the moisture in use by the timber industry. For our purposes, the useful value is the DM expressed as % of the WW, hence it is necessary to program the instrument to display the said value. To avoid any doubt, it is important to summarize the meaning of each of the conventions defined previously:

1. Moisture in % of the WW:

$$M[\%] = \frac{WW - DW}{WW}$$

2. DM in % of the WW (the useful value for our scopes):

$$DM[\%] = \frac{DW}{WW}$$

3. ATRO moisture (% moisture referred to the DM):

$$M_{ATRO}[\%] = \frac{WW - DW}{DW}$$

4. DMATRO (% of WW referred to DM):

$$DM_{ATRO}[\%] = \frac{WW}{DW}$$

3.2.1.2 Wrong Setting of the Drying Temperature

As a rule, measuring the DM of feedstock for biogas production (typically silage, manure, etc.) requires a slow drying process, possibly setting the temperature at 105°C. A quick drying process, at high temperature, may bring the sample's water to boil, scattering small particles out of the sample dish, biasing thus the test's result. When measuring liquid substrates (e.g., sludge and digestate), the surface will dry first, forming a watertight layer that prevents the evaporation of the remaining moisture. To prevent the said phenomenon, adding a certain quantity of sand to the sample will both help the heat conduction and break the superficial tension of the liquid.

N.B.: The moisture analyzer measures just the weight difference, so the sand's additional weight poses no problem: just add it to the dish, press the tare button to bring the reading to zero, and then add the liquid sample to test.

3.2.1.3 Wrong Installation of the Moisture Analyzer

Like in any other electronic instrument, the moisture analyzer's readings are subject to thermal drift—room temperature changes can produce variations of the readings. It is of outmost importance to place the moisture analyzer in a place without air currents, away from heaters, radiators, or air conditioners. It is a good practice to turn on the instrument at least 15 min before starting the measure, so as to stabilize its internal temperature. In general, all moisture analyzers include either a precision weight for manual calibration, or a self-calibration function. It is highly advisable to calibrate the instrument, especially during extreme temperature seasons or after turning off the laboratory's heating or air conditioning (as during the night or weekends).

3.2.1.4 Measuring the DM of Silage

The ensiling or silaging process is a special type of fermentation, similar to the preparation of *Sauerkraut* (a.k.a. *choucroute* or *pickled cabbage*), that converts the sugars contained in the fresh vegetal matter into volatile fatty acids (VFA), mostly in lactic acid. If the silage is kept away from air, the lactic acid acts as a preservative, ensuring its nutritional properties. Lactic acid has a boiling point near to that of water; hence, it evaporates when measuring the DM of silage. Since lactic acid is a direct precursor of methane, it is not correct to account it as water when drying the biomass to obtain the DM, and successively the VS, because it is indeed VS itself. Since VS is the denominator in the BMP's formula, underestimating the VS translates in a wrongly high BMP value. The solution to avoid the evaporation of the lactic acid (and any VFA in general) when drying the silage to measure its DM, consists in adding a known amount of some hydroxide, usually caustic soda (NaOH) or spent lime [$Ca(OH)_2$]. This will form sodium or calcium lactate, a salt that does not evaporate when drying, but volatilizes at 550°C. The difference between the DW and the net ash weight (total ash weight minus the hydroxide added to the wet silage) is then the correct VS value. This technique does not work well with sugar beet and sweet sorghum silage, because the sugars contained in the said biomass ferment into alcohol, which does not react with NaOH and will evaporate anyway. A more complete, but quite more complicated method, consists in distilling the silage in a still, kept at near-boiling-point temperature (105°C) until dry, then collecting the condensate and measuring both the dry mass that remained in the still and the chemical oxygen demand (COD; sum of all volatile compounds) collected in the condensate. The method was described by Porter and Murray (2001).

3.2.1.5 Interpretation of the Scale's Technical Sheet

The error margin of the DM determination is usually specified in the technical sheet or in the manual provided by the moisture analyzer's manufacturer. It ranges from ±0.015% (upper range models) to ±0.25% (low-cost models). Some manufacturers provide two error values: one in the case where the sample is 10 g (maximum accuracy) and another in the case of 2-g samples (minimum test time, but lower accuracy). In this particular case, "%" is the measure unit of DM, so for a technical sheet reporting, for instance, the specification "±0.15%" actually states that the absolute error in the DM determination is ±0.15%. In this particular case, the sign "%" must not be confounded with the relative error (conventionally expressed as a percentage of the instrument's measure limit) or with the instrument's sensitivity (a.k.a. resolution, i.e., the minimum difference of weight that the scale is capable of measuring).

Example: The technical sheet of a moisture analyzer states a maximum error of 0.05% when measuring moisture with 10-g wet samples. Suppose we test a 12-g sample of wet biomass, obtaining 32.53% DM. According to the technical specification sheet, the true DM value could then be any value in the range of 32.48%–32.58%. The correct way to express the DM in this example would be 32.5%, because the centesimal figures are not relevant—we only know for sure that the DM value is bigger than 32% and smaller than 33%.

3.2.2 Measurement of the DM with an Oven and a Scale

Carrying out this test requires a lower cost of investment compared to the automatic moisture analyzer: it is enough to have a small electric oven or a hot air stove, setting the temperature at 105°C–110°C, and a scale with at least ±0.1 g accuracy, like the ones employed for weighing letters. Compared to the automatic measure by means of a moisture analyzer, weighing the wet sample, then drying it in the oven and weighing it periodically until no weight variation is detected, requires more manual labor, and increases the possibility of human errors. In Chapter 2, Section 2.1.2, Example 1 shows how it is possible to measure with good accuracy the DM, even if employing a domestic kitchen scale. From the said example we can deduce that the lower the scale's accuracy, the bigger will be the sample size necessary to obtain an acceptable result; hence, higher the energy cost and longer the time required to carry out the test. According to the Standard Methods, it is necessary to weigh the sample at regular intervals until no weight variation is observed. Since most biomasses are hygroscopic, the sample should be weighed while still hot (but the heat can be transmitted to the scale's weight sensor, altering the reading), or left to cool in a glass bell filled with silica gel, to prevent it from absorbing moisture from the air. A more comfortable, but energetically more expensive method consists in leaving the sample in the oven overnight, and checking the weight in the morning with short intervals. In general, a whole night in an electric oven is enough to dry a 200-g biomass or liquid sample, while 4 h may be enough when employing a forced convection oven. Summarizing, the procedure to measure the DM of any feedstock is the following:

1. Calculate the sample quantity as a function of the scale's accuracy [see Chapter 2, Section 2.1.2 (Example 1) and Table 3.1 in the next paragraph of this chapter].
2. Place the sample in the oven, previously heated to 105°C, and leave it long enough for its complete drying.
3. Weigh the sample while still hot or leave it to cool in an airtight container filled with silica gel, to prevent the absorption of moisture from the air.
4. Calculate the DM as % of the WW with the usual formula:

$$DM[\%] = \frac{DW}{WW}$$

3.2.3 Measuring the Ash Content and VS

Measuring the ash content and consequently the VS requires inevitably a scale with the best resolution possible, because ashes are usually less than 10% of the WW, often a fraction ranging from 2% to 5%. The standard procedure to measure ashes is the following:

1. Determine the DM with any of the two methods described in Section 3.2.1 or 3.2.2.
2. Set the muffle oven at 550°C and wait until the temperature has stabilized (depending on the muffle's control fineness, in general, when reaching 550°C, the temperature will cycle around the said value).
3. Place the dry sample in the hot muffle. Check that the smoke flows freely from the muffle's fume hole or chimney.
4. Leave the sample at 550°C until its complete calcination, typically 4 h are enough. The dish's bottom must be covered with white–gray ash (if char rests or embers are observed, place the sample back in the muffle for a few additional hours).
5. Weigh the sample while still hot or let it cool in an airtight container filled with silica gel, to avoid moisture absorption or hydrate formation.
6. Calculate the VS as follows:

$$VS[\%] = \frac{DW - Ash}{WW}$$

7. Check the error in VS determination, because the said value will be useful for the BMP test. The procedure to check the error propagation is the following:

Applying the theorem of the error of the quotient between two measured magnitudes to the definition of VS presented in point (f), we obtain the following:

$$e_{VS} = e_{(DW-Ash)} + e_{WW}$$

The first term can be further developed as follows:

$$e_{(DW-Ash)} = \frac{E_{DW} + E_{Ash}}{DW - Ash} = \frac{2E_{scale}}{DW - Ash}$$

The second term can be simply calculated with the following:

$$e_{WW} = \frac{E_{scale}}{WW}$$

Hence, the total error in the VS determination results to be

$$e_{VS} = \frac{2E_{scale}}{DW - Ash} + \frac{E_{scale}}{WW}$$

Practical Rule
The following practical rule is a consequence of these formulas.

In the case of "normal" substrates (silage, dung, vegetal waste, etc.) containing around 30% DM and 3% ash (referred to the WW), the sample size necessary to determine the VS with a given admissible maximum error (e.g., 1%) can be calculated as a function of the available scale's maximum error, with the following formulas:

$$e_{VS} = 0.01 = \frac{2E_{scale}}{0.3 \cdot WW - 0.03 \cdot WW} + \frac{E_{scale}}{WW}$$

$$0.01 = \frac{2E_{scale}}{(0.3 - 0.03) \cdot WW} + \frac{E_{scale}}{WW}$$

$$WW = \frac{2E_{scale}}{0.27 \cdot 0.01} + \frac{E_{scale}}{0.01}$$

$$WW = 740 \cdot E_{scale} + 100 \cdot E_{scale} = 840 \cdot E_{scale}$$

For instance, if the available scale has a maximum error of 0.1 g and we want to determine the VS of silage with 1% maximum error, the sample size

TABLE 3.1

Quantity of Sample Necessary to Keep the Error of the Calculated VS below 1%, as a Function of the Absolute Error of the Scale, E_{scale}, and of the Estimated Moisture of the Substrate

Substrate under Analysis	Substrate's Approximate Moisture	Minimum Quantity of Sample to Keep $e_{VS} < 1\%$
Dry: straw, meals, oilseed cake	Up to 40%	$400 \cdot E_{scale}$
Semidry: silage, grass, municipal waste, fresh manure	40%–80%	$1000 \cdot E_{scale}$
Wet: digestate, sludge	90%–95%	$3900 \cdot E_{scale}$

necessary to dry and successively to calcine, will be at least 84 g. If a moisture analyzer having 0.005 g absolute error (a frequent value for medium quality instruments) is employed instead, then a 4.2-g sample will be enough. Since in general, the specification sheets of moisture analyzers require an initial sample size of 10 g, it is possible to conclude that the same sample, once dry, can be calcined and the weight difference with the ash will represent the VS, without additional error.

Table 3.1, based on the practical rule deduced earlier, is useful as a guideline to calculate the quantity of sample necessary to measure the VS as a function of the absolute error of the available scale and of the estimated moisture of the material to analyze.

3.3 Measuring the COD and Total N

There are two ways to carry out these kind of tests: analytical methods and spectrophotometric methods. The first ones, in spite of not being particularly complicated, require much labor and clumsy tooling, but on the other hand allow testing both liquid and solid substrates.

The spectrophotometer is suitable for testing liquid substrates without macroscopic solids. Its use in industrial biogas plants is hence limited to such substrates like whey, sludge (filtered), olive mill wastewater, digestate, and similar ones.

Particular attention must be paid to the maximum value that the instrument is capable of measuring (a.k.a. top of scale). The top of the scale varies from one manufacturer to the other, and in some cases, even has three different scales in a single instrument—for instance, COD LR (low range, from 0 to 150 mg/l), COD MR (medium range, from 0 to 1500 mg/l), and COD HR (high range, from 0 to 15,000 mg/l). In the biogas world instead, sludge and liquid substrates usually have COD concentrations far beyond 40,000 mg/l.

To utilize the spectrophotometer, it will be hence necessary to dilute the substrate with distilled water, then aspire with the automatic micropipette the required diluted sample amount, then measure the COD, and finally multiply the instrument's reading by the dilution factor.

Practical Example

It is necessary to measure the COD of cheese whey with a spectrophotometer model HI 86814. According to the manufacturer, the error margin in the COD HR scale is ±220 mg/l.

The instrument's top of the scale is 15,000 mg/l. From the literature, it is known that the COD of whey is in the range of 40,000–80,000. The error of the burette employed to sample the whey is ±0.1 ml, and the error of the cylinder where the dilution will be made is ±0.5 ml. Define the test protocol and calculate the maximum error.

To facilitate the calculations and to obtain at least the order of magnitude of the sample's COD, it is advisable to perform a first test with a dilution factor equal to 10. In this way, the COD value of pure whey will result from multiplying the instrument's reading by 10. The whole procedure is as follows:

1. Dilute 10 ml of whey in 90 ml of distilled water, mix well, and sample the necessary quantity with the instrument's automatic micropipette (200 μl in this example).

2. Add the prescribed quantity to the vial, and follow the procedure defined by the manufacturer in the instrument's manual (usually incubate at 120°C during 2 h, mix gently, cool down, and finally measure with the instrument, but this may vary from model to model).

3. In our example, the measured value was 6880 mg/l, hence

$$COD_{real} = Dilution \cdot COD_{measured} = 10 \cdot 6{,}880 = 68{,}800 \left[mg/l \right]$$

4. The error of this measure will be the sum of the single errors—the error of the spectrophotometer plus the error of the graduated cylinder employed to prepare the diluted sample. Hence

$$e_{COD} = \frac{220 \left[mg/l \right]}{6880 \left[mg/l \right]} = 3.2\%$$

$$e_{dilution} = \frac{0.1 \, ml}{10 \, ml} + \frac{0.5 \, ml}{100 \, ml} = 1.5\%$$

$$e_{total} = 3.2\% + 1.5\% = 4.7\% \approx 5\%$$

5. Now that we know with good approximation, the COD of the substrate in question, two alternatives are possible: to repeat the test

with a dilution factor equal to 7, so as to obtain a measured value nearer to the real value, but with a smaller error margin, or keep this measure as the good one, accepting ±5% uncertainty.

If the liquid substrate under test contains macroscopic solids, like in the case of liquid manure or digestate, the procedure becomes more complex. First it is necessary to take a sample that can be assumed to be "representative" of the whole lot (for instance, by thoroughly mixing before taking the sample, in such a way that the liquid is as homogeneous as possible). It is necessary to filter the sample with a sieve (it will be enough with a stainless steel or plastic kitchen sieve, similar to the ones employed for sieving flour, or even those for sieving pasta or rice). The solids separated in this way must be analyzed with the muffle oven, so as to determine their VS. The liquid fraction must be diluted and its COD measured as already described in the earlier example. The total COD will be, with good approximation, the result of the following formula:

$$COD_{tot} = COD_{l.f.} + 1.3333 \cdot VS_{s.f.}$$

COD calculated in the said way tends to be slightly underestimated.

It should be decided case by case (depending on the solids content of the liquid substrate) if it is convenient to measure the COD as described, or if it is easier to evaporate the sample until completely dry and then to determine the VS.

Measuring N (total and/or ammonia) with a spectrophotometer is analogous to measuring COD. As a general procedure, it is necessary to check the measure range of the instrument, then find out the probable N content from the literature, and finally decide whether to dilute the sample or not. The eventual presence of solids is not a problem, because in most cases more than 90% of the N remains in the liquid fraction. It is furthermore possible to wash the filtered solid fraction with a given amount of distilled water, and then to measure its N content, because nitrates and ammonia are very soluble and hence the total error will be further reduced. For routine controls, the method described is good enough. Should it be necessary to determine the N content in nitrogen-rich semisolid substrates (chicken dung, swine manure, etc.), then it is advisable to take a sample, test half of it with our instruments, and send the other half to a specialized laboratory. With the cost of a single test, we will then have a concrete measure of the accuracy of our own field method, and will be able to correct our measures, because the difference between our values and those of the specialized laboratory can be considered as a systematic error. In general, our measures in the field will be underestimated, because the spectrophotometer cannot measure the N contained in the solid fraction. Hence, the comparison of two tests made on the same sample, one in the field with the approximate method and the other performed by a specialized laboratory (assumed as errorless) will give us the correction factor to apply to future tests.

3.4 How to Design a Biological Experiment

The scope of biological tests is measuring the quantity and chemical composition of the biogas (or the net quantity of methane) produced by a given quantity of substrate, under predefined reference conditions, by actually digesting the substrate with a suitable inoculum.

In this section, we will explain the necessary data and procedure required to carry out a biological experiment. The following information is necessary before launching any biological test:

1. The VS concentration (or the COD) of the substrate that must be tested. The said value is an indicator of the presumably digestible fraction. In general, it is possible to have an approximate idea of the order of magnitude of the substrate's BMP (at least for usual substrates); hence, on the basis of the estimated yield, it is possible to define the amount of substrate to be employed in the test. For instance, suppose that the available instrument is an eudiometer with 250 ml capacity. Suppose that corn silage must be tested, which presumably will produce between 350 and 500 ml/g SV. Hence, the maximum amount of VS that can be employed is 0.5 g, otherwise it will be necessary to vent from time to time some of the biogas during the whole test period, usually 30 days, which in turn will increase the total error.

2. *The VS concentration of the inoculum*: This datum provides an idea of the quantity of active microorganisms capable of degrading the substrate. The VS concentration of the inoculum is of fundamental importance, since it will influence directly the error propagation.

3. The technical features of both the available reactors and gas measure system. It is necessary to have a clear idea of the reactors' volumes, error margin of the gas measure device, measure range (of flow, or volume, or pressure of the produced gas). It is important to be clear if the device is measuring gross biogas or net methane. In the second case, the error or the gas composition analyzer will add to the other instrumental errors.

4. *The scope of the test*: Do we need to check the health state of our plant, or the BMP of a given organic matter? It is necessary to define clearly what information to obtain.

5. *The representativeness of the sample*: Some substrates (e.g., municipal organic waste) are extremely heterogeneous. It is impossible to sample a full truck of municipal waste, homogenize its content, and take from the resulting mass the few grams of sample required for our tests. In such cases, there are two alternatives to measure the BMP:

a. To employ a big reactor or pilot plant, capable of receiving many kilograms of feedstock each day (assuming that the said amount is sampled randomly in such a way that it can be considered "representative"), or

b. To perform a large number of independent laboratory tests, whose results will be analyzed in statistical terms (average and dispersion of the BMP values, seasonal trends, etc.). With the same criterion, if the scope of the test is measuring the BMP of liquid manure from a stall, it will be necessary to mix thoroughly the content of the stall's catch basin, so as to avoid sampling just the surface scum, or the sediment, and to obtain a sample as homogeneous as possible.

Once all the above questions are clear, it will be possible to start the test's preparation, no matter which test it will be. The preparation phases, common to all anaerobic biological tests, will be explained in the next paragraphs. The descriptions are in general valid for any kind of instrument, unless contrarily stated.

3.5 The Preparation of Both Inoculum and Sample

As a rule, the inoculum for our tests will always be a certain amount of sludge taken from the digester (in the case of single-stage biogas plants). If the AD plant consists of several stages, the inoculum must be sampled from different tanks, depending on the scope of the test:

1. The inoculum for the SMA test (described in Section 3.10.5), must be sampled from the main digester.

2. The inoculum for testing the fermentative activity must be sampled from the prefermenter or from the first stage.

3. If the scope is to measure the BMP of a given substrate, it is possible either to employ a mixture of sludge samples taken from each reactor, or to employ digestate extracted from the last stage. The advantage of the digestate is having a smaller residual BMP than active sludge, and thus shorter preincubation times and less error of the result. The influence of the "background noise" introduced by the inoculum will be explained in detail in the next chapter.

4. To measure the residual BMP (i.e., the conversion efficiency of the plant), the sample must be taken from the digestate tank or postfermentor.

Depending on the literature consulted, it is possible to appreciate that two different schools of thought exist about how to prepare the inoculum. The first one, with roots in the industry of wastewater treatment, states that the inoculum should be taken from a plant already in stable operation, washed, centrifuged, and resuspended in water. Table 3.2 shows the norms that adopt the said criterion. The resulting VS concentration is usually comprised between 1 and 5 g/l.

The second school of thought, with roots in the agricultural and agro-industrial biogas industry, focuses on the potential production of energy by means of AD and not on the degradation of a given pollutant. Nevertheless, for the said scope, there is no universally recognized standard, although several have been proposed, as for instance, the German norm VDI 4630/2006 *Fermentation of organic materials—Characterization of the substrate, sampling, collection of material data, and fermentation tests*, which was updated in 2014. The first version of 2006 is available in German and English, but the version 2014, as to May 2017, is only available in German. The said norm provides guidelines both for tests aimed to the optimization of biogas plants and for tests to assess the greenhouse gas potential of waste for environmental studies, but does not consider the error propagation issues (it states just the maximum acceptable dispersion between samples). The inoculum should be employed "as it is" (no medium or any other substance that can be considered "booster" must be added) and its activity should be eventually checked by means of reference substrates (sodium acetate and cellulose). Many operators consider that the procedures required by VDI 4630/2014 are too complex for industrial plant laboratories to comply.

TABLE 3.2

International norms on the anaerobic degradability of some organic pollutants, most of them applicable to wastewater treatment

Body–Nr. (Year)	Title
ECETOC no. 28 (1988)	Guideline for screening of chemicals for anaerobic biodegradability
ISO 11734 (1995) EN ISO 11734 (1998) BS EN ISO 11734 (1999) UNI EN ISO 11734:2004	Water Quality—Evaluation of the ultimate anaerobic biodegradability of organic compounds in digested sludge— Method by measuring biogas production
ASTM E2170 (2001)	Standard test method for determining anaerobic biodegradation potential of organic chemicals under methanogenic conditions
OECD 311 (2006)	Anaerobic biodegradability of organic compounds in digested sludge by measurement of gas production
ASTM D5210—92 (2007)	Standard test method for determining the anaerobic biodegradation of plastic materials in the presence of municipal sewage sludge
ISO 15985:2004	Plastics—Evaluation of the ultimate anaerobic degradation in conditions of high solids content AD— Method based on the analysis of the produced biogas.
ISO 15473:2002	Soil quality—Guidance on laboratory testing for biodegradation of organic chemicals in soil under anaerobic conditions

Another proposal, cited at the end of this chapter, was presented by Angelidaki et al. in a congress of *International Waste Association* (IWA). According to the authors (in line with many other researchers), inoculums taken from agricultural AD plants (energy crops or agro-industrial byproducts) *should not* be washed and resuspended. Any pretreatment on the inoculum should be limited to separate eventual inert matter (wood chips, silt, small stones) but the rests of vegetal fiber must not be eliminated, because they serve as support for the microorganisms to grow; hence, they are useful to maintain the biodiversity of the bacterial ecosystem.

As of December 2016, the author has been involved in the redaction of the Italian norm on BMP batch assay for industrial AD scopes. Testing the greenhouse gas potential of biomass is out of the norm's scope. The draft of norm considers a simplified assay, consisting of a blank reactor filled with only inoculum and two sample replicates. The inoculum should be sieved with a 5-mm mesh and its VS content must be less than 5%, otherwise it should be diluted. Medium must be added to each reactor so as to ensure that an eventual shortage of microelements does not bias the result. It is admissible to pretreat the substrate in any way, but such pretreatment must be clearly stated in the test report. The inoculum must be preincubated for at least 5 days, and its bacterial activity should be checked periodically (three or four times a year), at least with sodium acetate and cellulose. The Italian draft of the norm includes an informative chapter on error propagation but it is not mandatory to include the error analysis in the test report. This is in net contrast with the ISO/IEC Directives, Part 2: *Principles and rules for the structure and drafting of ISO and IEC documents*. The said directives state that a norm on laboratory test methods must establish the acceptable instruments, their calibration protocol, the analysis or the test's accuracy, and even the rules for rounding the results.

It is thus clear that neither the VDI4630 nor the future Italian UNI norm complies with the ISO directives, so there is much margin to improve the accuracy of anaerobic biological tests.

The main features and flaws of the VDI 4630, the IWA proposal of norm, and the Italian draft of norm are treated in detail in Chapter 6.

An important aspect that the AD plant manager must consider is the inoculum's *specificity*. Inoculums employed by university laboratories are usually prepared from sludge and digestate from several plants, and hence they tend to be very biodiverse bacterial ecosystems, capable of quickly adapting to any kind of substrate. Sometimes, it is observed that BMP values published in scientific congresses and papers tend to be higher than those effectively encountered during the operation of AD plants or in one's own laboratory. One of the many reasons is the inoculum's specificity or ability to digest a given substrate. A common situation found in practice is the case of agricultural biogas plants. Figure 3.1 shows the result of a real experiment on pure cellulose, using two different inoculums. The first one was sludge from an anaerobic sewage treatment plant, while the second was sludge from an agricultural digester usually fed with corn silage and bovine manure.

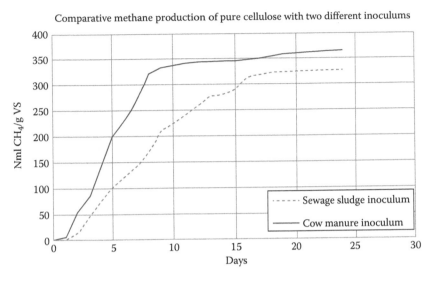

FIGURE 3.1
Digestion of pure cellulose with two different inoculums. Since ruminants have specific cellulolytic bacteria in their guts, the inoculum from the biogas plant fed with bovine manure shows a quicker and complete degradation of the substrate. Humans (and monogastric animals in general) do not possess such bacteria, so the degradation of cellulose with sludge from the sewage treatment plant was slower and incomplete.

It is then clear that the bacterial ecosystem of the sludge from an agricultural biogas plant will be quite adapted to digest short chain polysaccharides and cellulose. Suppose that in a given moment, the plant's manager has the possibility of getting free, or at very low cost, some kind of animal waste containing proteins and fats, for instance, slaughterhouse waste or residual glycerol from the biodiesel industry. According to scientific literature, such kind of feedstock has a high BMP, but in the author's experience, feeding it directly to an agricultural digester may cause some trouble. Since the bacterial ecosystem is not adapted to degrade high concentrations of protein and fats, the most probable effect will be to slow down the plant's productivity and in some extreme cases even bring it to the biological collapse. Sampling sludge from the digester and testing it in the lab, for instance, with the slaughterhouse waste, will produce a given BMP. Repeating the test with the same substrate, using the digestate from the first test as inoculum, will most probably yield a higher BMP. The reason is that the trophic relationships between the different species composing the bacterial flora will evolve so as to adapt to the new substrate, but once adapted (usually after two or more tests) it will be fit to digest both cellulosic and protein-based substrates. The shape of the degradation curve, fitting more or less to the classical sigmoid, will show to the experienced user if an inoculum is suitable for digesting a given substrate or not.

Figure 3.2 shows the results of an experiment aimed at determining the anaerobic degradability of seaweeds by two different inoculums: cow manure and anaerobic sludge sampled from the seabed. It is quite evident that the marine sludge is capable of digesting the substrate from the first day, but it exhausts the said capability after a short time, being the resulting BMP lower than the one reached by the bovine manure. The latter, on the contrary, did not begin to produce appreciable quantities of methane until 4 days, because the bovine bacterial flora had to adapt to digest algal biomass, an unusual substrate for it. Nevertheless, the bigger biodiversity of the bovine manure led to reach a higher BMP value than the one yielded by the marine sludge. The reason is that the algal biomass had more chance of being degraded by the large number of bacterial species present in cow manure while the marine sludge, in spite of containing bacteria already adapted to strive in high salinity and complex algal polysaccharides, was biologically poorer and hence less efficient. This experiment shows how an inoculum, initially unsuitable to degrade a given substrate, can become suitable with time.

At this point, it seems clear that, if planning to change the diet of the digester, it is the plant manager's responsibility to decide how to prepare the inoculum and decide the scopes of the test. For instance, if the manager plans to feed the biogas plant with a substrate having a very different chemical composition from that of the usual feedstock, then it is necessary to check if the said substrate is suitable or potentially harmful for the bacterial ecosystem. It may be necessary to find out how to adapt the existing inoculum to digest the new substrate.

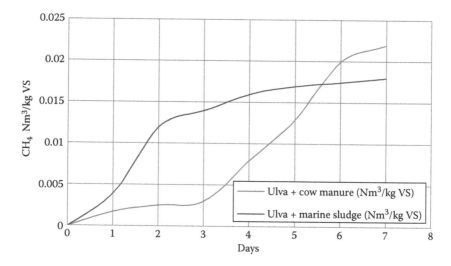

FIGURE 3.2

Anaerobic degradation of *Ulva* sp. algal biomass using marine bottom sludge and bovine manure as inoculum.

The following are some of the possible strategies that can be adopted:

1. Obtain digestate considered as "optimum" from a plant that is already regularly fed with the substrate in question and run a comparative test with inoculum from your own plant (difficult but not impossible).

2. If planning to restart a plant after an inactivity period, mix a sample of its sludge with sludge from the "donor" plant (the active plant that will supply the inoculum for the restart of the inactive one). Mixtures with different proportions of both inoculums can help to determine the minimum amount of external inoculum necessary for a successful restart, minimizing transport costs.

3. Adapt the plant's own inoculum to digest the new substrate. This is the easiest choice, but may require longer time if the initial biodiversity of the inoculum is too poor. In such case, it is advisable to take the inoculum from the plant already in operation, and test the new envisaged substrate mixed in different proportions with the currently employed feedstock. Eventually, the test can be repeated employing the digestate resulting from the first test as inoculum for a second test with increased proportion of new substrate. The goal of this method is to find out the maximum amount of new substrate that the inoculum is able to digest without losing its methanogenic activity.

Finally, it is necessary to consider the preincubation of the inoculum before starting the test. The said operation, called *degassing*, is necessary to exhaust the degradable organic matter already present when sampling the inoculum, so that the measured VS represent with good approximation the mass of living bacteria. Furthermore, a degassed inoculum will produce less *background noise* (the volume of gas produced by the inoculum itself). Degassing the inoculum sensibly reduces the error of the BMP assay and correlated tests. To degas the inoculum, follow the procedure described here:

1. Sample sludge from the digester, sieve it with a 5-mm mesh, measure its VS, and incubate it at the temperature chosen for the test (usually the same temperature at which the plant is running).

2. Observe the daily gas production. The inoculum is "degassed" when the cumulated production curve tends to remain flat for at least a few days, or when the daily gas production is less than a certain value, which varies according to different norms and protocols. The author considers that a good criterion to establish if the inoculum is already degassed, is to check that the daily net methane production is less than 1% of the total cumulated volume since the beginning of the degassing procedure, and remains so for at least 3 days.

Another criterion is to check that the specific daily production (the total methane production in 1 day divided by the total VS in the reactor) is less than 5 Nml/g VS.

3. As indicative values, the degassing procedure may take from 1 week in the case of sludge from a sewage treatment plant to 30 days when fresh cow manure or sludge from a plant running on energy crops is employed.

3.6 The Inoculum/Substrate Ratio, I/S

When performing biological tests, it is important to define the ratio between the "live weight" of the microorganisms and the "available food" for them. An acceptable measure of the live weight is the VS concentration of the inoculum after sieving and degassing it, as already explained in Section 3.5. The VS of the substrate provides a good measure of the "presumably biodegradable" organic matter in it. The typical ratio between the inoculum's VS and the substrate's VS for batch tests ranges from 1.6 to 3, so as to ensure that the quantity of living bacteria will be enough to consume all the substrate. Values smaller than 2 sometimes may lead to the phenomenon known as *inhibition from substrate* (in practice, bacteria can suffer from indigestion because of excess of food, or because the said food is too heavy to digest). I/S ratio values higher than 3 present the risk that the amount of gas produced by the substrate is of the same order of magnitude of the inoculum's background gas production. The said situation is likely to arise if the inoculum is not completely degassed and/or if the substrate is hardly degradable (e.g., animal feathers and hair), and the result will be an amplification of the instrumental error of the whole test.

Convention: In the scientific literature, it is frequent to find the inverse definition, S/I ratio, and consequently, the resulting ratio values will be smaller than 1. Throughout this text, the conventional definition adopted by the author has been I/S, hence practical values ranging from 2 to 3.

Practical rule: Always perform your tests with I/S=3.

Once the VS of both inoculum and substrate are known, it is necessary to calculate how much of each to load in the lab reactors. For this calculation, it is usually assumed that the densities of both inoculum and substrate are equal to 1 (in general, fairly true for the inoculum, but not always true for the substrate). The said convention simplifies the calculation and induces a negligible error.

The following example shows the calculation procedure.

It is intended to calculate the BMP of a substrate having 32% VS (on WW). The inoculum that has been taken from the digester, once degassed and sieved, contains 7% of VS on WW.

Suppose that the available reactors are 500-ml bottles and that the test will be performed with a volumetric measure device. It is necessary to leave a reasonable space to collect the gas, called "head space." If the test is to be carried with a barometric method, the said head space will typically range between 50% and 60% of the bottle's volume. If a volumetric method is employed, the head space should range between 15% and 20% of the total volume. Suppose that in the former example, a volumetric measure system will be employed, and that the head volume of the reactor is arbitrarily chosen to be 20% of 500 ml. Consequently, the total volume occupied by the mixture (inoculum+substrate) will be 400 ml. On the basis of the considerations already exposed in the former paragraphs, and on the practical rule, the I/S ratio adopted for this test will be equal to 3. It is hence possible to express mathematically the test conditions with the following formula:

$$\frac{Q_{inoc}}{Q_{subst}} = 3 = \frac{V_{inoc} \cdot VS_{inoc}}{V_{subst} \cdot VS_{subst}}$$

Furthermore

$$V_{inoc} + V_{subst} = 400 \text{ ml}$$

It is hence necessary to solve a system of two equations with two variables (i.e., the volumes of inoculum and substrate, respectively). Replacing V_{subst} in the first equation by its equivalent, deduced from the second one, we obtain

$$3 = \frac{V_{inoc} \cdot VS_{inoc}}{(400 \text{ ml} - V_{inoc}) \cdot VS_{subst}}$$

$$3 \cdot (400 \text{ ml} - V_{inoc}) \cdot VS_{subst} = V_{inoc} \cdot VS_{inoc}$$

$$3 \cdot 400 \text{ ml} \cdot VS_{subst} = V_{inoc} \cdot VS_{inoc} + 3 \cdot V_{inoc} \cdot VS_{subst}$$

$$\frac{3 \cdot 400 \text{ ml} \cdot VS_{subst}}{VS_{inoc} + 3 \cdot VS_{subst}} = V_{inoc}$$

By replacing the VS values, we obtain

$$\frac{3 \cdot 400 \text{ ml} \cdot 0.32}{0.07 + 3 \cdot 0.32} = V_{inoc} = 373 \text{ ml}$$

Even if it is not fully true that the density of both substrate and inoculum is equal to 1, the said assumption is fairly true for the inoculum and fluid substrates. Since the I/S ratio is measured in g, pouring 373 ml of inoculum on 27 g of substrate will produce a mixture with the desired I/S ratio, and its volume will be very close to 400 ml.

3.7 Defining the Mix Ratio (Intensity of Stirring)

Stirring is a parameter of fundamental importance for most industrial digesters. Too much stirring "stresses" the microorganisms, and too little stirring makes it difficult for the bacteria to easily access the nutrients. The calculation method for correctly setting the ideal intensity of stirring given the plant's size will be explained in Chapter 4. As far as laboratory reactors are concerned, the only easy thing to do is to procure by any means that they have at least a minimum of stirring. It has already been shown in the former chapter that many of the methods for measuring the gas amount produced by AD do not foresee stirring, mainly because of constructive reasons, but mostly for historical tradition, because most of them were initially conceived for testing wastewaters. As a matter of fact, when testing substrates with low solids content (<5% VS), and if the said solids are smaller than 1 mm, the difference between results obtained with static reactors and those obtained with stirred reactors will be negligible. The explanation is that the convective motion caused by the reactor's heating, together with the nutrients being in dissolved phase or in suspension, allow the bacteria to easily assimilate the nutrients at any time. In the case of agricultural or municipal waste plants, where the sludge usually contains from 8% to 12% VS (in some special digesters, known as "semidry," it can even reach 20%), it is necessary to induce some sort of mechanical stirring which must be also applied to laboratory reactors working in the same conditions. Some laboratory instrument manufacturers propose magnetic rod stirrers, which are just magnets coated with plastic, placed in the bottom of the reactor, which in turn is placed on a plate (sometimes equipped with an electric heater too) that generates a rotating magnetic field, forcing the magnetic bar to forcing the magnetic bar to spin inside the reactor. Such solution, once again, is borrowed from the industry of wastewater analysis, but is not suitable for testing agricultural inoculums and substrates, because they are quite viscous and contain fibrous solids, both hampering the magnetic bar's ability to rotate. Reactors having an electric motor and a gearbox connected to a shaft with paddles, or a bent rod, or helical propellers, are quite more efficient. Tests carried out with such reactors will be more accurate and their results closer to real-life plant operation.

Figures 3.3 and 3.4 show a simple batch reactor with just 1-l capacity and stirring rod, and a "professional" model with 5-l capacity for continuous tests.

Laboratory operators and many scientific publications too adopt the stirrer's speed as the main criterion for comparisons between tests, but just because most laboratory stirrers have a revolutions per minute or an "RPM" scale. For AD plant optimization, the concept "stirrer's speed" is not useful at all. In other words, adopting a single RPM value for all tests means for sure to define an internal standard of the laboratory in an arbitrary way, being then difficult to extrapolate the test results to a different instrument with a different kind of stirrer, not to say, to the real-scale digester itself. The stirrer's

FIGURE 3.3
One-liter batch reactor made of glass, with electric motor and gearbox driving a bent rod. (Photo Courtesy of Bioprocess Control AB.)

speed is an inadequate criterion to define the necessary stirring intensity for a given reactor because the stirring intensity is not just a function of the speed, but also of the stirrer's shape, dimensions, Reynolds number at which it is operating, and finally of the reactor's geometry. It may sound strange, but the digester's temperature plays an important role too, because the viscosity of the sludge depends, among other factors, on the temperature.

The *Unit Power* is a very interesting criterion for defining a reactor's stirring intensity, because it is more rational and easier to apply when the stirrer is driven by a DC motor with gear reduction. This criterion, originally developed in the USA, consists on calculating the stirrer's set point upon easy-to-measure physical variables. The *Unit Power* is based on an empirical fact: the AD processes work well when the power at the shaft of the stirrer (regardless of them being propellers or paddles) is comprised between 5 and 8 W/m^3 of digestion

FIGURE 3.4
Five-liter continuous reactor made of stainless steel with mechanical stirring provided by a brushless motor and electronic drive. (Photo Courtesy of Bioprocess Control AB.)

volume. At the laboratory scale, this means 6–8 mW/l. It is not easy to measure the power at the shaft of a very small motor with gearbox, like the ones usually employed for laboratory reactors. The correct way of measuring the mechanical power would be to connect a dynamometer and a tachymeter to the stirring shaft, so as to measure the torque and rotation speed, the power at the shaft being just the product of said magnitudes. Unfortunately, it is difficult to find in the market suitable instruments for measuring very small torque and it is also difficult to adapt them to existing reactors. Since this kind of test is sporadically needed in industrial anaerobic plants, buying a tachymeter and dynamometer is a superfluous expense. Fortunately, quite often the stirrers are driven by DC motors with permanent magnets, like the one shown in Figure 3.3. *Only in such special case,* it is possible to estimate, with good approximation, the mechanical power at the motor's (or gearbox's) shaft. The procedure is the following:

1. With the motor disconnected from its power source, measure with a tester the resistance across its terminals, R.

2. Mount the stirrer on an empty reactor. Connect the motor to its DC power source and let it run idle. Measure with the tester the voltage at the motor's terminals and the current. The idle DC power, W_0 (W), is the product of the voltage, U_0 (V), by the current, I_0 (A). The value W_0 represents all the energy losses (both electrical and mechanical). In the case of small reactors (<2l), in general, the voltage will be of the order of some V, but the current will be of the order of a few mA, hence the product will be the power expressed in mW. The power dissipated as heat in the windings, called $W\Omega$, is calculated as the product of the resistance R by the square of the current. Hence, the mechanical power when turning idle, called W_{Mo}, represents all the frictional losses, both of the motor and its gearbox. W_{Mo} is calculated as the difference between W_0 and $W\Omega$

$$W_{Mo} = W_0 - W_\Omega = U_0 \cdot I_0 - I_0^2 \cdot R$$

3. Now repeat the test with the same stirrer, but this time with the reactor full of sludge. Since sludge is a non-Newtonian fluid, the power absorbed by the motor will vary with time until reaching a steady state of temperature and apparent viscosity of the sludge, usually after some minutes. Hence, it is necessary to keep the motor stirring the sludge during at least 2 or 3 min before measuring voltage and current. Thus, we will be sure of measuring the actual stirring power in steady state. If in doubt about when reaching the steady state, it is possible to connect the amperometer since the beginning and wait until the current value is reasonably stable. At this point, the current (I) flowing through the motor's windings in steady state, dissipates power as heat in the resistance (R) of the windings. If the voltage drop along the connection cables and internal resistance of the DC power source is negligible, then the voltage at the motor's terminals, U, should be the same as U_0 measured in step 1 (if it is not, clean well the connectors and if possible use thicker cables or a power source with bigger rated power). Under the present operation with the reactor full of sludge, the current, I, will be higher than the one measured with the empty reactor, I_0. We will call W_t the total electrical power absorbed by the motor under load condition, and W_R the power dissipated as heat by the windings' resistance, calculated as the product of the square of the current I and the measured resistance R:

$$W_t = U \cdot I$$

$$W_R = I^2 \cdot R$$

4. Calculate the mechanical power dissipated by the motor's (or gearbox's) shaft on the sludge, W_m, as the difference between the total

power under load, minus the frictional losses measured under idle condition, minus the heat dissipated across the windings' resistance under load. Hence

$$W_m = W_t - W_{Mo} - W_R$$

5. Finally, the quotient between the power W_m and the volume of sludge in the reactor, V, must be a value comprised between 0.005 and 0.008 kW/m³ (or between 5 and 8 mW/l, which is the same). In general, the said quotient will be higher than the ideal values, so it will be necessary to define *ON/OFF* cycles in such a way that, in average, the mechanical power at the shaft of the stirrer lies within the specified interval. Translated in formulas

$$5\,[\text{mW/l}] < \frac{W_m \cdot t_{ON}}{V \cdot (t_{ON} + t_{OFF})} < 8\,[\text{mW/l}]$$

6. Choosing t_{ON} and t_{OFF} is arbitrary. Nevertheless, we should consider another factor that helps increasing the useful life of the stirrer motors: the startup current. At each start, the current across the motor's windings can be a peak of several times the nominal current. Such peak overheats the windings, but if t_{ON} is long enough, the heat will have enough time to dissipate. Very frequent starts and stops do not allow enough time to dissipate the heat generated by the start current, hence, it is a good practice to assume $t_{ON}+t_{OFF}$ equals to 3600s (1h), so as to grant enough time for the motor to cool down before the next start.

Example

Assume we desire to set the stirring rate of a reactor similar to the one shown in Figure 3.3. The digestion volume is 400 ml. With the empty reactor, we measure the following voltage and current:

$$V_0 = 8.8\,[V]$$

$$I_0 = 42\,[\text{mA}] = 0.042\,[A]$$

Furthermore, the resistance of the windings is

$$R = 19\,[\Omega]$$

Hence

$$W_0 = 370\,[\text{mW}] = 0.370\,[W]$$

$$W_\Omega = I_0^2 \cdot R = 0.042^2 \cdot 19 = 0.0335\,[W]$$

so

$$W_{Mo} = 370 - 33.5 = 336.5 \, [mW]$$

Now we fill the reactor with the sludge and leave the stirrer running at full load for at least 2 min, at the temperature chosen for the biological test. After the mentioned time, we measure the following voltage and current:

$$V_t = 8 \, [V]$$

$$I_t = 71 \, [mA] = 0.071 \, [A]$$

Hence

$$W_t = 0.568 \, [W] = 568 \, [mW]$$

and

$$W_R = I^2 \cdot R = 0.071^2 \cdot 22.4 = 0.113 \, [W] = 113 \, [mW]$$

The power dissipated on the sludge (the *Unit Power*) is therefore

$$UP = \frac{(568 - 336.5 - 113) \, [mW]}{0.4 \, [l]} = 296 \, mW/l$$

The given value is 37 times bigger than the one prescribed for the *Unit Power*, so stirring 97 s every hour (or 48 s every half hour) is the minimum stirring acceptable. It is not convenient to exceed the calculated *Unit Power*, because increasing the time the motors remain on only increases their wear, but does not change the biogas production.

Remark 1

The former example, taken from a real case (AMPTS II with 5 V DC motors and 30:1 gearbox), allows to calculate the efficiency of the DC motor and gearbox, as follows:

$$\eta_{motor+gear} = \frac{W_t - W_{Mo} - W_R}{W_t} = 20.9 \, \%.$$

Such a low efficiency is normal in very small DC motors with high gear ratio as the one considered.

Remark 2

The error of the UP calculated as described above is rather high, and will depend on the error class of the instrument employed. If a good quality instrument is not available, then it is advisable to perform several tests on different stirrer motors and assume an average value.

3.8 The BMP Test: One General Procedure, Multiple Applications

Now that we know how to obtain the basic data (VS of the substrate and inoculum) and how to plan a biological test, we will study in detail the BMP test. In extreme synthesis, the scope of the BMP assay is measuring the net methane quantity produced by any combination of inoculum and substrate—or a mixture of substrates—under given test conditions. The BMP assay has multiple applications:

- Checking the effective methane yield of any biomass, mimicking as close as possible the operational conditions of our biogas plant (i.e., with the actual bacterial ecosystem, at the same temperature and with the same nominal HRT of our biogas plant). The scope in this case is to define the digester's diet and/or to check the purchase price of the feedstock.

- Checking the residual methane quantity that our biogas plant's digestate can still produce. The scope of this test is to measure the efficiency of the feedstock's conversion into energy.

- Measuring the health state of the bacterial ecosystem (the procedure is explained in detail in Section 3.10).

- Testing different operational strategies (different temperatures, different stirring rates, different particle sizes of the substrate, and others that will be described in detail in Chapter 4). The scope is deciding the most suitable management strategy of the biogas plant.

- Testing pretreatments (e.g., adding enzymes to silage) or commercial additives. The scope is evaluating the effectiveness of the pretreatment or additive and checking if the eventual improvements pay back their higher operational costs.

3.8.1 General Procedure for Measuring the BMP

1. Degas the inoculum as already explained in Section 3.5. If employing digestate as inoculum, this operation helps to measure the biogas plant's conversion efficiency (see Chapter 4).

2. Prepare the mixture of substrate and inoculum in the correct I/S ratio, as already described in Section 3.6.

3. Fill at least two (better three) reactors with the said mixture, close them hermetically, connect them to the gas measure device, and finally purge the head volume with an inert gas, so as to eliminate any trace of oxygen. These reactors are called *sample replicates* or just *replicates*.

4. Fill at least one reactor (better two or three) with only the degassed inoculum. The said reactor(s) is (are) called *blank*.

5. Adjust the stirring intensity according to the guidelines in Section 3.7.

6. Start the experiment. If employing manual reading instruments, take at least one reading per day, *always at the same time*.

7. Copy the reading in a spreadsheet. If employing instruments that do not have the capability of normalizing the volume in real time, then a barometer and an ambient thermometer are necessary too. Copy in the spreadsheet the readings of ambient P and T, which will be necessary to normalize the gas volume reading with the formulas explained in Section 2.2.1.2, Chapter 2.

8. If employing a CO_2 trap (a bottle filled with NaOH solution) between the reactor and the measure device, the gas volume reading represents the net amount of methane. If employing a barometric method or a volumetric method without CO_2 trap, then it is necessary to measure the composition of the biogas with any suitable method available. *Do not assume that the biogas is always 60% methane and 40% CO_2, because such convention is seldom true.* Copy the measured composition of the biogas in the spreadsheet.

9. At the end of the experiment (conventionally 30 days), make a backup copy of the spreadsheet and elaborate the data as explained in Section 3.9.

3.9 Processing the Measured Data

At this point, we will have a table with "raw data." In the most favorable case (instrument with data logger and real-time normalization of the gas volume), we will have a table containing the normalized gas volumes with a temporal axis divided in constant intervals (typically hours or days). In the worst case (eudiometer or similar instruments), we will have a table containing the daily volume of gas produced by the reactors, together with the ambient P and T readings, and finally the chemical composition of the gas, if pertinent.

3.9.1 Step-by-Step Data Processing

1. Start from a spreadsheet containing the following columns: day (0, 1, 2, 3, ..., N), daily gas production (methane or biogas, according to the case) of the replicates no. 1, no. 2, etc., idem of the blank(s), ambient P and T at the moment of each reading.

2. In the first cell of a separate column, type the following formula: average of the replicates' readings minus average of the blanks, such

difference divided by the VS of substrate added to each reactor. Copy down until the last row, corresponding to day N.

3. In the column aside the former one, normalize the gas volumes by means of the formulas explained in Chapter 2, Section 2.2.1.1.

4. Plot the normalized values as a curve, employing the graph function of the spreadsheet.

5. Check the error propagation according to the procedure explained in Section 3.11. The said operation is not mandatory, but it is anyway advisable, so as to evaluate the reliability of the measure. This is especially necessary when the results are doubtful (too low or too high) or when the replicates yielded gas amounts very different from each other. In the latter case, such differences could be the result of a hardware problem (reactors not completely air tight, problems with the measure instrument), or may be the consequence of a human error during the preparation of the experiment (wrong quantities of substrate, substrate too heterogeneous or not enough homogenized, some kind of contamination of the inoculum, etc.). The comparison between the error of the calculated average BMP and the dispersion between single replicates allows a finer diagnostic of the eventual problems. For example, if the error of the calculated average BMP is in the order of 5%, but the dispersion of the single replicates is bigger than said value, then the cause is most probably a problem with the instruments or a human error in the preparation of the test, as will be explained in the following Section 3.9.2.

3.9.2 Frequent Causes of Problems during the BMP Test

3.9.2.1 Problems Caused by Human Errors

Carrying out BMP tests is easy and, when it becomes a routine, quite often the operator works with the "automatic pilot." Distractions unavoidably lead to errors (of weighing, in the preliminary calculations for the experiment preparation and so on). Figure 3.5 shows an example from real life. It was planned to measure the BMP of sorghum and corn with an AMPTS II. For several reasons, it was necessary to start immediately with the test and there was no time to measure the VS *before* starting. Hence, the mixture for the replicates was prepared assuming VS from the literature, defining I/S=4, so as to be sure that the actual ratio was at least 3. The test was started; the VS were measured during the next day, finding 16% for sorghum and 24% for corn. In the case of the sorghum sample (dotted lines), the VS quantity effectively loaded in the reactors was 6.6 g (40 g WW), and the resulting I/S was equal to 5. In the case of the corn sample (continuous lines), the VS were 8.9 g (37 g WW), resulting in an I/S ratio equal to 3.7.

The instrument employed for this test, a second generation AMPTS, has caustic soda CO_2 traps and normalizes the volume in real time. Therefore,

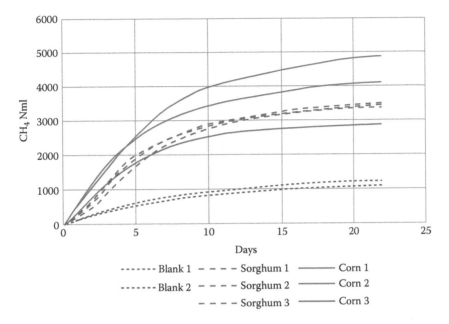

FIGURE 3.5
Curves of normalized methane production of each single reactor.

the curves shown represent the normalized production of methane of each single reactor. Observe that the curves of both the blanks and the sorghum are almost coincident while the curves of the corn samples are quite disperse. Since the particle size of the corn sample was a bit coarse, a certain dispersion had to be expected. Such kind of uncertainty is related to the representativeness of the sample rather than to the instruments or the method, but a big dispersion was not expected. Observe that the curves *corn 1* and *corn 2* are almost coincident until the 5th day, and hence their small dispersion can be attributed to the nonuniformity of the sample and its coarse trituration. The curve *corn 3* is suspiciously flat, although it follows the same shape of the other two. If a problem had been caused by the reactor (e.g., gas losses), or by the instrument (defects or problems in the sensor of the measurement cell, obstruction of the gas tube, etc.), the curve would be either interrupted or its shape would show "steps" or other forms of discontinuity. In this case, it is quite easy to suspect that the discrepancies between curves can be caused by a human error—the test was prepared in a hurry, with the scope of teaching some students, so most probably the reason of the anomalous curve *corn 3* is simply that there was some mistake in the preparation of the sample. This example is useful to justify why employing triplicate samples is always a good practice. Some protocols, like the German VDI 4630/2014, define that the number or replicates must be *at least* 3. The Italian draft of norm considers acceptable at least 2 replicates, so as to reduce the overall cost of each test.

Observe that if the Italian protocol had been applied in the preparation of this experiment, such a big dispersion between both curves would have led to discard the test and repeat it. Since triplicate samples were employed and two of them gave coherent results, one can simply discard the results of the sample *corn 3* and calculate the BMP with the average of the remaining ones. Therefore, the result with the average of *corn 1* and *corn 2*, divided by the VS in each reactor is

$$\text{BMP}_{\text{corn}} = \frac{0.5 \cdot [(4103 + 4881) - (1238 + 1112)]}{8.9} = 373 \, \text{Nml/g VS}$$

As a mental exercise, consider the said BMP as the base to calculate the VS that were effectively loaded in the reactor *corn 3*:

$$VS_{\text{presumed}} = \frac{2873 \, \text{Nml}}{373 \, \text{Nml/g VS}} = 2.4 \, \text{g}$$

As can be observed, the WW of a sample containing such amount of VS is 9 g. This means that the most probable cause of the anomalous curve *corn 3* is that the student mistook the quantity of VS with the quantity of fresh matter that had to be loaded in the reactor. A typical example of human error.

3.9.2.2 Problems Caused by Instrumental or Method Errors

Sometimes, it may happen that a reactor loosely closed, or a tube accidentally disconnected, causes gas leaks or even the entry of air, blocking the process. Such is the case depicted in Figure 3.6., which is the result of some tests for measuring the BMP of *Arundo donax* (common cane, Castilian cane).

The reactor *Arundo2* had to be moved during the test, so most probably one of its tubes was accidentally loosened, leaking biogas and allowing air to enter. In this case, the problem is easy to solve because the three curves were almost coincident until the reactor was moved on the 18th day of the test. We can just calculate the BMP_{30} considering the average of reactors *Arundo1* and *Arundo3*.

Practical Conclusion No. 1
It is advisable to perform always the tests by triplicate. Instrumental problems can always happen and even the most experienced laboratory operators, working with top quality instruments, are not exempt.

Practical Conclusion No. 2
Once you start a test, do not move the reactors or the measure device until your experiment is finished, so as to avoid gas leaks.

We will conclude this paragraph with an example on the errors originating in the work method. An undergraduate student was assigned the task of measuring the BMP of some agro-industrial waste (presumably containing a

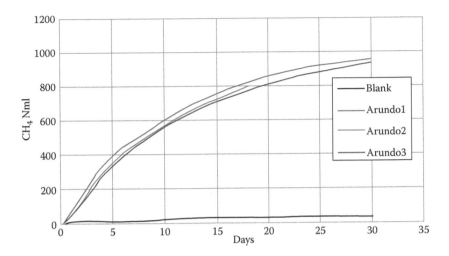

FIGURE 3.6

Curves showing the normalized production of methane from fresh *A. donax*. Observe the interruption of the AD process in the reactor called *Arundo2*.

certain percentage of proteins and fats). The student decided to carry out his tests with I/S=2, because it is the most frequent test condition in *the literature* (absolutely true, but questionable). The inoculum employed was sampled from a biogas plant running with cow manure and corn silage. Figure 3.7 shows the result of the test.

It is evident that such test is wrong, and in this case, it is not necessary to investigate the causes like a detective. We can discard instrumental problems because the test was performed with a brand new AMPTS II, perfectly calibrated and checked before the start. Gas leaks can be excluded too because both curves are coincident and interrupted at the same time. By analyzing the method adopted by the student, one finds out the following errors:

1. General test planning: if one suspects that a substrate may contain fats, it is advisable to perform the test with I/S ≥ 3, so as to avoid provoking the "indigestion" of the bacteria.

2. The inoculum had not been preincubated, as can be easily deduced from the large amount of gas produced by the blank and the shape of the blank's curve. A well-preincubated inoculum should yield a very flat and almost linear curve.

3. The test was carried in duplicate, which in this case increases the uncertainty.

4. The inoculum was sampled from a biogas plant that was usually running on cow manure and corn silage, i.e., high carbohydrate content and low protein and fats. The substrate to test was presumably

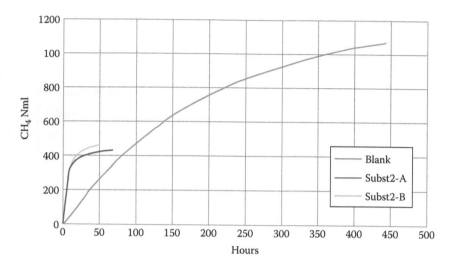

FIGURE 3.7
Example of a test failed because of method errors.

composed of the inverse mix: little carbohydrates and much protein and fats. Therefore, even in the supposition that the aforementioned factors (1) and (2) had been correctly planned, the inoculum lacked anyway of specificity. A two-step test would have been necessary, employing the digestate of the first test as inoculum for the second, so as to allow the bacterial flora to adapt to the new feedstock.

5. The quantity of inoculum loaded in the blank (400 ml) was different from the one loaded in the replicates (150 ml). Even if the test had not been interrupted by the inhibition of the inoculum, a direct calculation of the BMP would have been impossible, requiring the inclusion of a correction factor and hence amplifying the errors in the calculation.

6. When checking the procedure for the determination of the VS, it was found that the student employed just 1 g of sample "because the laboratory has a precision scale." The DW turned out to be 0.250 g, and the ash 0.055 g. Checking the scale's manual, it resulted that the sensitivity of the instrument was 0.005 g but its accuracy was 0.02%. Remember that, by international convention, the accuracy is always expressed as a percentage of the end of scale (110 g for this concrete model). Consequently, the absolute error of said scale is 0.02% of 110 g, i.e., ±0.022 g. Calculating the error of the VS with the formula already explained in Section 3.2.3 (repeated here for convenience):

$$e_{VS} = \frac{2 \cdot E_{scale}}{DW - Ash} + \frac{E_{scale}}{WW},$$

we obtain the following results:

$$e_{VS} = \frac{2 \cdot 0.022}{0.250 - 0.055} + \frac{0.022}{1} = \pm 25\,\%$$

When determining the VS of the inoculum, the student did a better job, but anyway the result was not acceptable. The WW was 5 g, resulting in just 0.380 g DW, while the ash was 0.035 g. Hence

$$e_{VS2} = \frac{2 \cdot 0.022}{0.380 - 0.035} + \frac{0.022}{5} = \pm 13\,\%$$

Therefore, the error in the determination of the I/S results to be

$$e_{I/S} = 25\,\% + 13\,\% = \pm 38\,\%$$

7. Such an error in determining the VS is already unacceptable, and furthermore it is amplified when calculating the I/S ratio. In practice, we cannot be sure which the I/S was, since any value comprised between 1.24 and 2.76 could be theoretically possible. We can reasonably conclude that the sequence of errors led to performing a test with an effective I/S < 2. Hence, the sudden interruption of the methane production after only 60 h resulted from the inhibition of the bacterial activity, caused simultaneously by the excess of substrate and by the scarce specificity of the bacterial flora to digest a very "heavy" substrate, composed mainly by protein and fats.

3.10 The Hydrolytic and Methanogenic Activity Tests: Checking the Bacteria's Health

The procedure described in the former Sections 3.8 and 3.9 applies to any substrate whose BMP is unknown. A particular case of the BMP test is measuring reference substrates, i.e., pure substances whose BMP is known in advance.

This test turns useful in comparing different instruments (assuming the inoculum is the same for both and adequate to the substrate). It is advisable as practice for students and beginners (easy to find out eventual problems in case the test is not correctly performed). In the case of the biogas plant manager, it is an important test to check if the different microbial communities are healthy and consistent in number. Recalling what was explained in Chapter 1, Figure 1.2 and Section 1.1.2.2, the degradation of the organic matter and its conversion in biogas is a sequential process. To test each link of the chain, there is a reference substrate for an AD assay (polysaccharides, proteins, lipids, and VFA). If the amount of methane obtained during the said tests is lower than a reference value, or even null, it means that the population of each specific bacterial group is inhibited or even missing.

Table 3.3 provides a list of the reference substrates and their range of acceptable BMP and the group of bacteria that can metabolize each one.

In the next paragraph, we will analyze, case by case, the peculiarities of each bacterial activity test compared to the generic procedure of the BMP test.

3.10.1 Degradation of Polysaccharides (Glucose, Starch, and Cellulose)

As shown in Table 3.3, these substances have all a reference BMP in the range of 350–400 Nml/g SV. The test is easy to perform because the said substrates are easily digestible. To save money, it is possible to employ industrial products available in any supermarket, instead of their "laboratory grade" equivalents: white sugar or glucose syrup; corn, potato, or rice starch; paper tissue or cotton. The reference BMP measured with such commercial substances may result in being closer to 350 Nml/g VS than to 400 Nml/g SV, because they are not 100% pure. It is enough to account for the purity grade, usually indicated in the label, as explained below:

- *Glucose*: Usually sold in supermarkets as a syrup, contains some water. Consequently, its theoretical BMP will be 373 Nml/g VS multiplied by the purity percentage declared in the label.

- *White sugar*: Mostly composed of sucrose (saccharose), which is a disaccharide (two glucose molecules bound together). It may contain

TABLE 3.3

Reference Substrates for Biological Activity Tests

Reference Substrate	BMP Range (from Literature)	Specific Group of Bacteria
Cellulose	352–410 Nml/g VS	Hydrolytic bacteria
Starch	370–410 Nml/g VS	Hydrolytic bacteria
Glucose	350–373 Nml/g VS	Acidogenic bacteria
Casein or gelatin	370–470 Nml/g VS	Proteolytic bacteria
Propionic acid	330–350 Nml/g COD	Acetogenic bacteria
Butyric acid	330–350 Nml/g COD	Acetogenic bacteria
Fatty acids (vegetable oil)	800–1000 Nml/g VS	Lipolytic bacteria
Acetic acid	330–350 Nml/g COD	Acetoclastic *archaea*
Sodium acetate (anhydrous)	354–373 Nml/g VS (330–350 Nml/g COD)	Acetoclastic *archaea*

some additive and moisture, but in general, it is almost pure, hence we can assume its theoretical BMP to be equal to that of pure glucose.

- *Fructose*: Same chemical composition of glucose, but it tends to absorb more moisture than white sugar, hence it is necessary to check thoroughly its TS and VS before calculating the quantity for each reactor.

- *Starch*: It is a polysaccharide (several glucose molecules bound together) and its chemical composition is quite variable. For example, corn starch available in supermarkets contains, according to the manufacturer, "87% carbohydrates," the rest practically being all moisture. Therefore, tests performed with this substrate should yield less than 357 Nml/g on WW. *Rice starch, instead, is usually available in the bank "baby care" or in the bank "diet products," being its purity in general more than 99%.* Being a hygroscopic substance, it is advisable to measure its TS each time, before starting the AD test.

- *Cellulose*: It is a polysaccharide, and its generic chemical formula is $(C_6H_{10}O_5)_n$. Pure microcrystalline cellulose (in powder) is available in specialized shops of laboratory materials and reagents. Industrial cotton (either as flocks or as disks) is composed of 99% cellulose. To increment its degradability, it is mandatory to chop finely its fibers with a scissor, this operation being easier with cotton in flocks. Ideally, the fibers should be shortened and broken up in such a way to avoid the formation of clumps when adding them to the inoculum. Tissue paper is instead composed of short fibers (to facilitate its disintegration in the septic tanks and sewage ducts). It is easier to employ than cotton, but its cellulose percentage is lower, about 80%–90%, depending on the quality. Furthermore, tissue paper can contain traces of Cl or other additives potentially inhibitors for the anaerobic process. Three-layers "soft rolls" are generally composed of 90% pure cellulose and their fibers are short enough for easy comminution, making this quality of tissue paper a good reference substrate for AD tests.

3.10.2 Casein and Gelatin in Powder

Both products are composed of pure proteins.

The elementary analysis of proteins in general shows the following composition: 55% carbon, 7% hydrogen, and 16% nitrogen. Proteins are extremely complex and heavy molecules, some of them having a molecular weight equal to 10,000. They all differ from each other by spatial structure, although their average elementary composition is quite similar to the values indicated.

Casein is extracted from milk whey. It is available in shops specialized in laboratory supplies and as food integrator in some shops of dietetic products for body builders and athletes. Gelatin in powder is available in almost all supermarkets. The one employed for the preparation of desserts usually contains sugar, colorants, and flavors, and hence it is not suitable for our scopes. Neutral gelatin in powder (the one employed for the preparation of galantine, aspic, and similar dishes) is the best alternative as reference substrate for AD tests, because it dissolves easily and can be dosed quite accurately in each reactor. The time required for anaerobic degradation of proteins is generally long (about 20 days) and not all inoculums are able to fully digest them ($BMP_{30} < 370$ Nml/g VS).

3.10.3 Propionic and Butyric Acids

Propionic acid ($C_3H_6O_2$) and butyric acid ($C_4H_8O_2$) are both liquids having a rather pungent smell (especially the second, which is the usual component with of rancid butter and vomit). Both are available only in specialized shops of laboratory supplies. Because it is well known that acids consume the alkalinity of the inoculum, it is a good practice to neutralize them either with caustic soda (NaOH) or with sodium bicarbonate ($NaHCO_3$). Consequently, the bacteria will digest sodium propionate or butyrate ($C_3H_5O_2Na$ or $C_4H_7O_2Na$). Sodium propionate is a chemical commodity, known in the food industry with the (European) code E281. It is a widely employed preservative for industrial bread, because of its anti-mold properties and is commercially available in specialized industrial food supplies stores. Sodium butyrate is an ingredient for pharmaceutical laboratories, since it is employed for the preparation of some perfumes and homeopathic medicines.

The theoretical BMP of both propionic and butyric acids are the same of acetic acid, i.e., 350 Nml/g COD. Sodium propionate yields 530 Nml CH_4/g VS, while sodium butyrate yields 515 Nml CH_4/g VS.

To calculate the I/S when employing any of both acids (and hence, it is impossible to determine the VS), the COD concentration of pure acid can be calculated as follows:

$$C_3H_6O_2 = 112 \text{ g COD}/74 \text{ g acid} = 1.51 \text{ g COD/g propionic acid}$$

$$C_4H_8O_2 = 160 \text{ g COD}/88 \text{ g acid} = 1.81 \text{ g COD/g butyric acid}$$

Performing the test

1. Filter the inoculum with a mesh (advised 5 mm, in any case smaller than 10 mm) so as to separate fibers and coarse solids.
2. Prepare a mixture of 0.5 g of propionic acid and 0.5 g of butyric acid, per liter of inoculum; hence, in total 1.66 g of total COD.
3. Prepare a blank reactor and a sample reactor loaded with the previously mentioned mixture. Start the experiment and check that the net production of methane is in the range of 435–581 Nml (350 Nml/g COD).

N.B.: In this test, the usual I/S criterion (I/S=2 or 3) is not valid, because the presence of propionic and butyric acid in proportions bigger that 1 g/l can inhibit the methanogenic *Archaea*. Furthermore, the inoculum could already contain a certain amount of both acids (especially if it is a "fresh," nondegassed inoculum). As a reference guide, the mixture of acids described here has 1.66 g of COD, while the filtered inoculum may contain between 30 and 40 g COD/l; hence, the I/S ratio for this concrete test is bigger than 18.

3.10.4 Fatty Acids (Lipids)

Lipids (oils and fats, either animal or vegetal) are very heavy molecules with more than 6 C atoms. Their high BMP has easily deceived many incautious biogas plant managers, with rather disastrous consequences—biologically collapsed biogas plant or at least heavy loss of productivity. It is necessary to know that such high BMP values, often found in the scientific literature, are seldom reachable in industrial biogas plants, especially if their HRT is less than 60 days. Furthermore, since lipids degrade slowly and are hydrophobic substances, they tend to form emulsions. Lipids present a certain inhibitory capacity for the inoculum. For instance, stearic and oleic acids, usual components of vegetable oils having 18 carbon atoms, are toxic for the AD process at concentrations as low as 1 g/l. A dose in the range of 80–130 mg/l produces already a certain inhibition.

The equivalence coefficients between VS and COD introduced in the former chapter are not valid for lipids. The reason is that lipids are composed mainly of carbon and hydrogen, so their total oxidation requires more oxygen than any other feedstock. For instance, consider oleic acid (main component of olive and sunflower oils). Its total oxidation will undergo the following reaction:

$$C_{18}H_{34}O_2 + 25.5\ O_2 \rightarrow 18\ CO_2 + 17\ H_2O$$

$$282\,g + 816\,g \rightarrow 792\,g + 306\,g$$

Hence, the amount of COD per g of oleic acid is given by the following proportion:

$$\frac{816\left[g_{COD}\right]}{282\left[g_{oleic}\right]} = 2.89$$

To test the capacity to digest lipids, the rule of thumb is to consider the COD of oils and fats as three times their DW, and to choose the I/S in terms of COD, considering the inoculum's COD equal to 1.333 its VS, expressed in g/l. With such rule, the final concentration of oil and fats will be smaller than their usual inhibition limits, i.e., 0.6 g/l. We must expect long times to observe an appreciable methane quantity in our gas flow meter. Furthermore, the shape of the curve will look more like a straight line than like a sigmoid.

Example: We desire to test the lipolytic capacity of a given inoculum that contains 5% VS. The reference substrate chosen is olive oil and for the sake of simplicity, its composition is assumed as 100% oleic acid.

The inoculum contains therefore a quantity of VS equal to 5% of 1000 g/l, hence 50 g/l. The VS concentration, expressed as COD, is then 1.333 times said value, i.e., 67 g/l.

Suppose we want to perform the test with 0.2 g of olive oil per liter of inoculum (toxicity limit of oleic acid according to Angelidaki and Ahring). Such dosage is equivalent to 0.6 g of COD of oleic acid, 50 g COD/l inoculum in total, and hence I/S=83.

Observe that, in general, oils and fats are not composed of pure oleic acid, but contain other fatty acids that are not so strong inhibitors of the AD process. As an example, Cirne et al. found that the lower toxicity threshold of triolein (a triglyceride, composing 75% of olive oil) becomes evident with I/S=2.2 (in terms of COD). Consequently, in our hypothetic test with olive oil, we could safely add up to 5.5 g of oil/l of inoculum [hence I/S=50/(5.5×3)=3].

Since the exact composition of animal and vegetable fats is always uncertain, as well as the specificity of the inoculum to digest them, it is advisable to perform several tests with increasing I/S ratios, e.g., 3, 5, 10, and 15, and see which ones suffer inhibition. Remember that the duration of such tests must be always at least 60 days, as demonstrated in Figure 3.8.

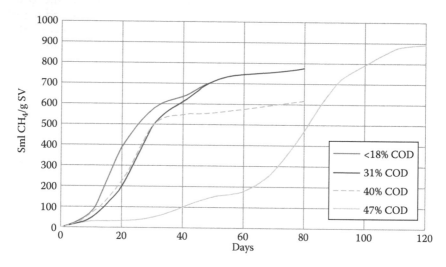

FIGURE 3.8

Degradation of triolein, according to Cirne et al. (2007). Graphics simplified by the author. According to the cited study, all tests where the amount of fat was less than 18% of total COD gave coincident curves. Tests with 31% of total COD suffered some inhibition, but at the end recovered and yielded substantially as much methane as the former. Tests with 40% of COD suffered inhibition and did not recover, yielding less methane than the former. Tests with 45% of total COD yielded the highest BMP, but took 120 days.

3.10.5 Acetic Acid and Sodium Acetate

Both substrates, direct precursors of methane by the acetoclastic group of *Archaea*, allow to carry out two complementary tests: The Specific Methanogenic Activity (SMA) test and the methanogenic capacity test. Both are useful to understand if the inoculum is apt or not, or if the biogas plant has a good resilience margin toward inhibiting agents, or if the bacterial eco-system is collapsed. The general philosophy of the SMA test assumes that, since the acetoclastic *Archaea* produce the biggest part of methane, adding acetic acid (or sodium acetate) to a given inoculum must result in a consistent production of methane. If not, it means that the inoculum contains no *Archaea*, or contains an irrelevant quantity, or that the *Archaea* are inhibited. Such a condition may arise when employing *aerobic* sludge from a sewage treatment plant, or in the case of diagnosing problems in a digester already suffering a biological collapse (process already inhibited by any cause).

The test consists in loading the batch reactor with inoculum and a given quantity of acetic acid, or sodium acetate, and checking if the BMP reaches a value close to the theoretical one. This first phase is called *methanogenic capacity test*, and the information it gives is whether the inoculum is capable of converting the acetate in methane, or not. The second phase, i.e., the *SMA test*, consists in assessing the maximum methane flow generated by the inoculum in the presence of acetic acid or sodium acetate, referred to the mass of presumably live and active bacteria (i.e., the VS of the inoculum).

From the same test, we will then be able to deduce two different kinds of information: qualitative and quantitative (the methanogenic capacity and SMA, respectively). The said information is necessary to decide if the last link of the trophic chain in the AD process (the methanogenesis) is as solid as the chain itself, or if it is the weakest step or bottleneck of the whole process. At this point, we need a short theoretical digression, before studying some practical examples.

3.10.5.1 Stoichiometry of Acetic Acid and Sodium Acetate

The anaerobic degradation of acetic acid can be described with the following chemical formulas:

$$CH_3COOH \quad \rightarrow \quad CH_4 \quad + \quad CO_2$$

$$60 \text{ g acetic acid} \quad 16 \text{ g} = 22.4 \text{ Nl} \quad + \quad 44 \text{ g} = 22.4 \text{ Nl}$$

Furthermore

$$60 \text{ g acetic acid} = 64 \text{ g COD}$$

then

$$BMP_{theoretical} = 22,400 \text{ Nml}/64 \text{ g COD} = 350 \text{ Nml/g COD}$$

The same reasoning applies also to sodium acetate

$$BMP_{theoretical} = 22,400 \ Nml/60 \ g \ VS = 373 \ Nml/g \ VS.$$

3.10.5.2 First Step: Checking the Methanogenic Capacity of Acetate or Acetic Acid

The test follows the same general procedure of the BMP test already explained. Special attention must be paid to the I/S ratio, because it has a direct influence on the measured value. Experimental evidence shows (Badshah et. al.) that employing sodium acetate as reference substrate with I/S < 3 will yield less methane than the theoretical production. The reason is the partial inhibition caused by the higher concentration of sodium. If employing acetic acid instead of sodium acetate, the results will be similar, in this case because of the partial inhibition caused by lowering the pH.

Figure 3.9 shows how the net methane production varies as a function of the I/S ratio when employing sodium acetate as reference substrate, according to Badshah et al. (graphics elaboration by the author).

Observe that the experimental error will be minimum when I/S=2–3, because the production of methane will be maximum, but at the same time the concentration of sodium will be maximum and hence the BMP could be smaller than the theoretical value because of the partial inhibition caused by the sodium. The test can last from 1 to 5 days.

By increasing the I/S ratio, the concentration of sodium drops and its inhibitory effect drops consequently. The duration of the test is usually

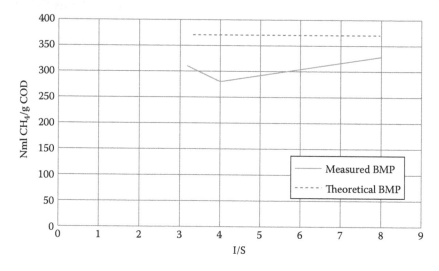

FIGURE 3.9
Variation of the BMP test results when employing sodium acetate, as a function of the I/S ratio. (Data from Badshah et al., graphics by the author.)

shorter, 2 or 3 days, but the experimental error of the test will increase, especially if the inoculum is too "fresh", i.e., insufficiently degassed.

As anticipated at the beginning of this paragraph, the test performed in the way described can be called "methanogenic capacity test," since it shows *if* the inoculum is capable of digesting acetate or not. It is not an SMA test, because it does not consider the time necessary for the complete degradation of the acetate.

The following protocol describes how to carry out the methanogenic capacity test with sodium acetate.

- Sample the inoculum from the digester and sieve it through a 5-mm mesh, so as to eliminate macroscopic fibers and solids.
- Measure the VS or the COD of the filtered inoculum (it is assumed that the measured VS represent the living bacterial biomass).
- Neutralize 6 g of acetic acid with 4 g of caustic soda diluted in 80 ml of distilled water. ATTENTION: This is an exothermal reaction. Wear gloves and eye protection.
- *Alternative*: Purchase anhydrous sodium acetate from a supplier of laboratory chemicals. Even if it is labeled "anhydrous," it adsorbs moisture once the pot has been opened. To reduce the error in the preparation of the test, bring the sodium acetate to the oven at 140°C and dehydrate it for 1 h.
- Complete with distilled water up to 100 ml (or weigh 10 g of sodium acetate as soon as you take it from the oven and dilute it in 100 ml of distilled water). The resulting solution will then have 64,000 mg/l COD.
- Prepare a sample reactor with I/S=5–8 and a blank reactor with the same quantity of degassed inoculum. Add distilled water until completing the same total volume in both reactors. Should the inoculum be "fresh," then either perform the test with I/S=3 or preincubate the inoculum to degas it.
- Check that the net methane production reaches 330–350 Nml/g COD within 2 or 3 days (if I/S > 5) or 5 days (if I/S=3). If it takes more time, it means that the inoculum has methanogenic activity, but it is inhibited.

It is possible to perform the former test with wine or apple vinegar instead of acetic acid, although the error will be higher because the exact acidity of such substrate is known only approximately. Follow the following procedure to prepare the test using vinegar as substrate:

Wine vinegar contains about 6% acetic acid; hence, it is necessary to add 4 g of caustic soda every 100 g of vinegar, so as to neutralize its pH. Start by adding and perfectly dissolving 3 g of caustic soda, then check with the pH meter that the resulting pH is less than 7 and add the remaining caustic soda in small quantities, checking the pH, until reaching at least 6.8, ideally 7. The resulting theoretical COD will be 6400 mg/0.1 l (i.e., 64,000 mg/l) and conversely, the resulting theoretical VS will be 6 g/0.11, i.e., 6%. Should you need more or less caustic soda for reaching the neutral pH, correct proportionally the reference value (64,000 mg/l). For instance, if the pH reaches 7 with just 3.8 g of caustic soda, then the COD will be

$$COD_{real} = 3.8 \text{ g}/4 \text{ g} \cdot 64,000 \text{ mg}/l = 60,800 \text{ mg}/l$$

and the corresponding VS will be

$$VS_{real} = 3.8 \text{ g}/4 \text{ g} \cdot 6\% = 5.7\%$$

3.10.5.3 Second Step: The SMA Calculation

In this case, the scope is not measuring *how much* methane the inoculum can produce from acetate, but rather *at what speed* the bacteria degrade the substrate. Hence, instead of focusing on the *total production* of methane (Nml/g COD or Nml/g VS), we will measure the *maximum daily flow per g of VS of the inoculum*. This is the reason why the test is called *"specific* methanogenic activity."

The test protocol is analogous to the one already described in Section 3.10.5.2. Adopt an I/S > 5, typically I/S=8 (if employing sodium acetate), otherwise employ acetic acid or vinegar, in such quantity as not to exceed 1 g acetic acid/l of mixture (adding the vinegar or acetic acid directly to the inoculum). Check daily the net quantity of methane produced. The said value must be divided by the *total VS of the inoculum*. In the case of agricultural digesters, the maximum quotient's value should be reached within 2 or 3 days, and must be comprised between 30 and 40 Nml/g $VS_{inoculum}$ d, with an "acceptable" minimum of 10 Nml/g $VS_{inoculum}$·d. The latter value in Nml/g $VS_{inoculum}$·d was deduced from Bolzonella (2013), who expresses the SMA in terms of COD. Other authors (Sandoval Lozano et al., 2009) report "usual" SMA values comprised between 17 and 70 Nml/g $VS_{inoculum}$·d (the said values are common in the case of anaerobic sewage sludge or industrial effluent treatment plants). In the case of granular sludge (typical of upflow anaerobic sludge blanket [UASB] or similar digesters), the SMA can reach values comprised between 280 and 1400 Nml/g $VS_{inoculum}$·d. In the extreme case of granular sludge, incubated at 60°C, the SMA reaches a peak at 2100 Nml/g $VS_{inoculum}$·d (Van Lier et al., 1995, already cited in Chapter 1, Figure 1.3). It must be noticed that in the wastewater treatment industry, many authors check the net methane flow on hourly basis, finding out the peak value and then multiplying it by 24 h, as a linear extrapolation. As an exercise, divide by 24 the indicative values of granular sludge quoted earlier—the corrected range results comprise between 12 and 58, very close to the range indicated by Sandoval Lozano et al. Whatever convention the reader will adopt, the important thing is to be coherent. Throughout this book, the author has adopted the first convention just because it was easier for him measuring the daily flow rather than the hourly flow.

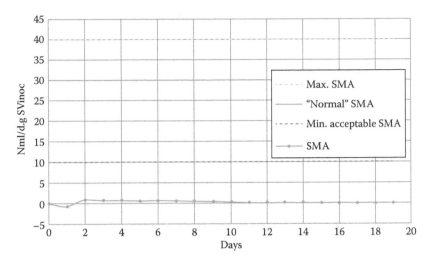

FIGURE 3.10
Example of inoculum with inadequate methanogenic activity, taken from a plant with problems. Results obtained with $I/S=5$ and $T=38°C$.

Important
In scientific literature, it is frequent to express the SMA in terms of COD CH_4/g $VS_{inoc}·d$. The reason is that, to perform the balance of COD, it is easier to employ such "scientific" unit. Nevertheless, laboratory instruments measure the volume of methane produced in a certain time, hence Nml/d. How to calculate the equivalence then? Remember that the quantity of methane produced by acetic acid or by sodium acetate (or by VFA in general) is 350 Nml/g COD (Section 3.10.5.2); hence, 1 g COD of $CH_4=350$ Nml CH_4.

Figure 3.10 shows the results of an SMA test performed with an inoculum that had an inadequate methanogenic activity, because it had been sampled from a digester having some biological problems. Observe the negative value the first day—this means that the blank produced more methane than the reactor with acetate, an undoubtable proof that the inoculum already suffered from acidification, so adding more acetic acid was enough for fully inhibiting it.

3.11 Analysis of the Error Propagation in the BMP Assay and Its Variants

At this point, we already know the measure techniques of the parameters that are relevant for the biogas plant's management, and the theoretical BMP values of the reference substrates when the inoculum is "healthy." We can finally calculate the uncertainty (a.k.a. error) percentage of our results, so as to check if we have performed our tests correctly.

By definition, the BMP of a given substrate is the result of the following formula:

$$\text{BMP} = \frac{(\bar{C} - \bar{B})}{\text{VS}},$$

where

$\bar{C}$ = average of the normalized CH_4 production of the sample reactors

$\bar{B}$ = average of the normalized CH_4 production of the blank reactors (ideally almost null if the inoculum has been adequately degassed).

VS = volatile solids of the substrate (it is assumed that all the sample reactors have been loaded with the same amount of substrate VS).

N.B.: The formula above is valid only if the quantity of inoculum in the blank and sample reactors is the same. It is possible to employ different quantities of inoculum in the sample and blank reactors, introducing a correction factor in the formula, but this increases the error. As a rule, we will always work with the same quantity of inoculum in the blank and sample reactors. If necessary, distilled water can be added to the inoculum in the blank reactors to have the same headspace volume in all reactors (necessary condition for minimizing the error induced by the gas moisture).

By applying the error propagation rules to the BMP formula, we get

$$e_{\text{BMP}} = \frac{e_{\text{instr}}(\bar{C} + \bar{B})}{(\bar{C} - \bar{B})} + e_{\text{VS}}$$

$$e_{\text{BMP}} \approx \frac{e_{\text{instr}}(\bar{C} + \bar{B})}{(\bar{C} - \bar{B})}$$

where

e_{instr} = error of the gas volume measurement instrument (including the normalization error if the instrument does not automatically normalize its reading)

e_{VS} = error in the determination of the VS, already explained in Section 3.2.3.

N.B.: The numerator of the first term in the given formula is valid only if the single quantities of CH_4 produced by the sample reactors are approximately equal, and hence close to their average. In such condition, it is possible to state that the absolute error of the average is equal to the relative error of the instrument, multiplied by the average itself. If the single values of the sample reactors should have different orders of magnitude, the results should be discarded, the causes of said dispersion of values should be found, and then the test repeated in the correct conditions. Otherwise, the result obtained with the average of very disperse single values would be unreliable, even if the calculated error may be acceptable.

Observe that, if the value $\bar{B}$ in the former formula was null (or anyway negligible compared to the value $\bar{C}$), then the first term will reduce to just e_{instr} (minimum possible error). This is the reason why it is very advisable (mandatory according to some norms and protocols) to degas the inoculum with some days of preincubation before starting the test. If the biogas plant is running correctly, then the digestate will contain almost no more nutrients available for the bacteria. Hence, if the said digestate is employed as inoculum, its residual biogas production (aka "background noise" or "background production" of the test) will be small. Consequently, the preincubation could be omitted, on condition of accepting a slightly higher general error of the test.

To calculate the second term of the BMP error formula it is then necessary to know the error of the instrument employed to measure the gas volume produced by the digestion. Table 3.4 shows the indicative errors of some instruments widely available in the market. Should the reader employ other instruments not listed here, he or she must take into account the maximum error specified by the manufacturer and the normalization error.

TABLE 3.4

Typical Measure Error Margins of Some Standard Instruments Diffused in the Market

Instrument	Max. Relative Error	Remarks
AMPTS II	$\pm 1\%$ (+ e_{ga} if NaOH filters are not employed)	Normalization in real time, resolution 10 ml, can measure net methane or total biogas
AMPTS light	$\pm 1\%$ (+ e_{ga} if NaOH filters are not employed)	Normalization in real time, resolution 10 ml, can measure net methane or total biogas
Gas endeavour	$\pm 1\%$ (+ e_{ga} if NaOH filters are not employed)	Normalization in real time, resolution 2 ml, can measure net methane or total biogas
μFlow Cell	$\pm 1\% + e_{ga}$	Normalization in real time, resolution 10 or 2 ml, generic gas flow meter
MilliGascounter Cell	$\pm (3\% + e_b + e_t + e_{ga}$ if NaOH filters are not employed)	6% = min. normalization error assuming constant T and P for the normalization formula
Eudiometer and similar water displacement instruments	$\pm (e_v + e_b + e_t)$	e_v = error of the graduated cylinder or burette, calculated as the quotient between the cylinder's error class and the reading
Oxytop Bottle	$\pm (1\% + 1\,hPa/P + e_{T} + e_v)$	e_v = error in measuring the head volume of the bottle e_T = error of the temperature of the gas in the reactor's head space P = measured absolute pressure
Hohenheim syringes	$\pm (5\% – 9\% + e_b + e_t)$	See an example of calculation in Chapter 6

e_b = error of the room barometer employed for the normalization of each reading
e_t = error of the room thermometer, idem
e_{ga} = error of the gas analyzer, including the uncertainty of the calibration mixture (when measuring total biogas production).

N.B.: The errors of barometers and thermometers are very variable from one model to another. Laboratory grade electronic barometers usually have 2% error while mercury column barometers usually have 1.5%. The accuracy of glass capillary thermometers is usually ±1°C, hence when measuring mesophilic temperatures, the relative error is in the range of 2%–3%. Electronic laboratory thermometers usually have 0.1°C–0.2°C absolute error; hence, their relative error when measuring mesophilic temperatures is in the order of 0.2%–0.3%. The error of gas analyzers is more difficult to assess, since it is a complex function of the error (tolerance) of the calibration mixture and the intrinsic measure error of the instrument. The error in the measure of the gas composition is usually in the range of 2%–3%.

3.12 A Controversial Technique: Correcting the pH, the Alkalinity, and Adding Nutrients before Starting the Batch Tests

Adding micronutrients to the inoculum is a widely diffused praxis in the scientific world, quite often reported in the literature and furthermore proposed by some authors as a standard procedure to follow at each test. The aqueous solution containing the minerals, nutrients, and other substances that favor the bacteria's growth is usually called *medium, nutrient broth,* or *mother solution* in the scientific literature. From a scientific point of view, adding a solution that contains all the buffer and nutrients required for an optimum digestion process has a sense when the scope of the test is finding the methane *potential* of a given biomass, i.e., the maximum amount of methane that the biological process is able to extract. If the goal is measuring the maximum value, it makes sense putting the bacteria in the best conditions to thrive. This means that the *medium* must contain all the mineral nutrients eventually lacking in both the substrate and the inoculum. In other words, the medium must provide enough buffer capacity to keep the pH as close as possible to 7, the organic nitrogen (biologically available, i.e., as nitrates and not in ammonia state), the micronutrients (P, K, Ca, Fe, Ni, and Co; according to some authors also traces of Se are necessary), vitamins, and other substances considered as probiotics for the microorganisms. In practice, the *medium's* formula should be decided in each specific case, according to the substrate and inoculum employed for the test. The German norm VDI4630/2014 states that no buffer and probiotics should be added to the inoculum, so as to measure the BMP of the substrate in the real operational conditions of the biogas plant. On the contrary, the Italian draft of norm (as to April 2017) includes a recipe of mother solution and the minimum amount of this to be added to the inoculum. The criterion adopted is to ensure that the microorganisms have at least a minimum amount of mineral nutrients, so as to avoid the underestimation of the BMP.

In the personal opinion of the author, both norms are wrong, or at least strongly biased by the personal background of the people who wrote them. The German norm is too much focused on agricultural biogas plants. In Germany, such biogas plants are fed mainly with cereals, and it is common practice to regularly add micronutrients to the digester so as to keep its bacterial ecosystem healthy. Hence, it is quite logical that adding micronutrients to the inoculum when performing a laboratory test—as advocated by the Italian draft of norm—is not necessary when the inoculum has been collected from such digesters, because it already has what it needs. In some cases, the addition of medium in the laboratory can be detrimental, because the excess of some minerals (Ni, Co, and Fe) will inhibit the hydrogenotrophic *Archaea*, resulting in lower methane yield. On the other hand, if the scope of the test is obtaining information for managing a biogas plant fed with sludge from sewage water treatment, quite often such inoculum will lack some essential mineral nutrient. It is then necessary to add medium to the inoculum before starting the laboratory test, so as to avoid the underestimation of the BMP.

From the practical point of view of the biogas plant manager, it is always advisable to perform a comparative test from time to time, so as to check if the addition of medium to the inoculum results in a higher methane yield, or if, on the contrary, it turns to be inhibitory. It is very important to check the error propagation of the test. Quite often, the author has read scientific papers where "improvements" of the methane yield in the order of 3% are presented as "positive results" demonstrating the efficacy of the micronutrients addition. We know for experience that it is hard to perform a BMP test with less than 3% error when employing very sophisticated volumetric instruments with real-time normalization, being 1% the minimum error possible with the current technology. On the contrary, 5% or more error margin is quite frequent when employing other instruments for the tests. When performing a comparative test, any difference of the same order of magnitude of the error margin is irrelevant, being impossible to state if the real cause of the difference is the addition of the nutrients or the instrumental errors.

Figure 3.11 shows an example from real life. Two reactors were filled with 1800 ml each of fresh inoculum, sampled from the first reactor of a three-step agricultural biogas plant. One reactor was left unaltered, while the second reactor was added with medium according to the Italian draft of norm. The difference between both was around 8% less methane in the case of the reactor that received additional micronutrients.

The former test was performed with an AMPTS Light, having $e < 1\%$. The difference encountered is bigger than the error margin of each individual methane production, and hence we can be sure that in this case the addition of medium to the reactors led to a partial inhibition because of micronutrients' excess. On the other hand, we cannot state with absolute certainty that the magnitude of the inhibition is 8%, because the difference between two measures, each one having 1% error, has its own uncertainty:

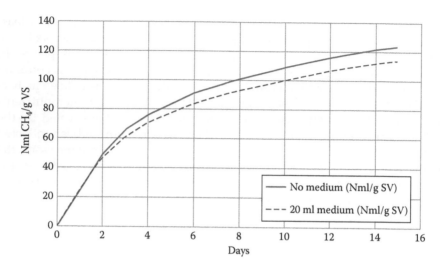

FIGURE 3.11
Negative influence of the excess of micronutrients in the inoculum.

$$E_{(S1-S2)} = E_{S1} + E_{S2} = 1\% \ (123 + 113) \mathrm{Nml/g \ VS} = 2.36 \ \mathrm{Nml/g \ VS}$$

Since the values of both measures are very close to each other, the relative error will be high:

$$e_{(S1-S2)} = E_{(S1-S2)} / (S1 - S2) = 2.36/10 = 24\%$$

In this case, 24% is the uncertainty of the calculated error.

In other words, we cannot be sure if the real difference between adding nutrients to the inoculum, or leaving it as it is, will be exactly 8%, since any value in the range of −6% to −10% could be possible. In any case, we can state for sure that the addition of medium to one of the reactors led to its partial inhibition, −8% being the "most probable" magnitude of said inhibition.

Bibliography

Angelidaki, I., Alves, M., Bolzonella, D., Borzacconi, L., Campos, J., Guwy, A., Kalyuzhnyi, S., Jenicek, P., and van Lier, J.B., Defining the biomethane potential (BMP) of solid organic wastes and energy crops: A proposed protocol for batch assays, *Water Science & Technology*, 59(5), 927–934, 2009.

Angelidaki, I., and Ahring, B., Effects of free long-chain fatty acids on thermophilic anaerobic digestion, *Applied Microbiology Biotechnology*, 37, 808–812, 1992.

Badshah, M., Liu, J., and Mathiasson, B., *The Biomethane Potential of Chemically Defined Substrates Evaluated Using Automatic Methane Potential Test System*, Lund University Publications, Lund, Sweden, 2012.

Bolzonella, D., *Monitoring and Lab Analysis—Assays*, Lectures on biogas technology, Jyvaskyla Summer School, Jyväskylä, Sweden, 2013.

Cirne, D.G., Paloumeta, X., Bjornsson, L., Alves, M.M., and Mattiasson, B., Anaerobic digestion of lipid-rich waste—Effects of lipid concentration, *Renewable Energy*, 32, 965–975, 2007.

Esposito, G., Frunzo, L., Liotta, F., Panico, A., and Pirozzi, F., Bio-methane potential tests to measure the biogas production from the digestion and co-digestion of complex organic substrates, *The Open Environmental Engineering Journal*, 5, 1–8, 2012.

Farina, R., Bernard, O., Steyer, J.P., Conte, T., Lema, J., Roca, E., Ruiz, G., Franco, A., Cellamare, C., Spagni, A., and Martinez, J.A., Misassumptions in COD measurement in wastewaters containing high concentration of VFA, ENEA publication, Protezione Unità Tecnica Valutazioni Ambientali—Laboratorio Protezione e Gestione della Risorsa Idrica, 2004.

Ficara, E., Scaglione, D., and Caffaz, S., Test di laboratorio per valutare la producibilità di biogas, Chapter 9. *Biogas da agrozootecnica e agroindustria*, Dario Flaccovio Editore, Palermo, Italy, 2011.

Porter, M.G., and Murray, R.S., The volatility of components of grass silage on oven drying and interrelationship between dry matter content estimated with different methods, *Grass Forage Science*, 56, 405–411, 2001.

Sandoval Lozano, C. et al., Microbiological characterization and specific methanogenic activity of anaerobic sludge used in urban solid waste treatment, *Waste Management*, 29, 704–711, 2009.

Schneiders, D., da Silva, J., Till, A., Lapa, K., and Pinheiro, A., Atividade methanegênica específica (AME) de lodos industriais provenientes do tratamento biológico aeróbio e anaeróbio, *Revista Ambiente & Água—An Interdisciplinary Journal of Applied Science*, 8(2), 135, 2013.

Sung, T., and Alastair, D., Long chain fatty acids degradation in anaerobic digester: Thermodynamic equilibrium consideration, *Process Biochemistry*, 45, 335–345, 2010.

VDI (Verein Deutscher Ingenieure), Norm VDI 4630—Fermentation of organic materials. Characterisation of the substrate, sampling, collection of material data, fermentation tests. Beuth Verlag GmbH, Berlin, 2006 (draft 2014 available only in German).

4

Application of Laboratory Experimental Results to the Management of the Biogas Plant

4.1 Practical Applications of the VS Test

The total amount of carbon, assumed *a priori* as "digestible," is exclusively contained in the solid fraction defined as volatile solids (VS). The VS test is very simple to carry out and provides much useful information for the supervision and control of the anaerobic digestion (AD) process.

4.1.1 Measuring the Organic Load (OL) and the Organic Load Rate (OLR)

The first practical application of the VS measure is checking the OL and the OLR, so as to assess that the plant is working within its design limits.

Most of the agricultural biogas plants built in Europe can be classified as the continuously stirred reactor tank (CSRT) type. This type of digester cannot accept an OLR >3 kg of VS/m³·day, otherwise the efficiency of conversion of the biomass into methane will drop to unacceptable levels. Typically, single-stage and two-stage biogas plants require the OLR to remain in the range of 2.5–3 kg of VS/m³·day. The so-called "compact digesters," promoted by some German manufacturers, can work with up to 5 of kg VS/m³·day. According to the same manufacturers, it is necessary to stabilize the bacterial ecosystem by adding a daily dose of "integrators" (mixtures of enzymes, minerals, and lyophilized bacteria) so as to keep the digestion efficiency within acceptable levels. The manufacturer's argument in favor of such digesters is that they do not belong to the CSRT type, but their working principle is that of the so-called plug flow digester (PFD), which is (theoretically) more efficient than the CSRT. In spite of the theoretical considerations and the peer-reviewed literature, the experience of some of the author's customers in Italy with such digesters has been very negative. The reasons are as follows:

1. It is not true that the commercial compact digesters are "real" PFRs. In practice, their operation is somehow in between those of a PFR and a CSRT.
2. The cost of the "integrators" is high, since they are proprietary formulas of the same digester manufacturer. Nevertheless, during the commercial operation of the plants the owners observed that the efficacy of such additives in keeping an acceptable digestion efficiency was much lower than the manufacturer's claims. Such low efficacy was demonstrated by batch tests in the laboratory too: the residual BMP of the digestate collected at the exit of the PFR was around 50% of that of the feedstock.

In the case of conventional single-stage CSRT digesters fed with agricultural biomass exceeding 3 kg of VS/m³·day can lead to the acidification and eventual biological collapse of the AD process. In the case of biogas plants with two or more digesters, the OLR must be calculated by dividing the daily OL by the sum of the volumes of all the digesters.

4.1.1.1 Practical Application

It is quite advisable to perform at least once a month the following tests:

1. Take a representative sample of the feedstock (or a representative sample mix, if the digester's diet consists of a mixture of different biomasses) and measure its VS. In the case of corn or other silage, a representative sample can be prepared by taking a handful of silage from at least six different points of the trench's loading front, thoroughly mixing them in a container and then taking the necessary amount of sample from such mixture. In the case of sludge or liquid manure, it is advisable to stir very well with a stick before taking the sample from the collection pit, and always avoiding to sample from the surface or from the bottom. Once you got a representative sample, measure its VS according to the ordinary procedure.
2. Calculate the OL with the usual formula:

$$OL = Q_{biomass} [ton/day] \cdot VS [\% \text{ of } WW]$$

3. Calculate the OLR with the usual formula:

$$OLR = \frac{OL}{\sum V_{digesters}}$$

Please note that, if your plant has an airtight storage tank for storing the digestate, its volume must be included in the formula above, since such a tank can be considered as a postdigester.

4. Take a sample of digestate from the storage tank (if it is of the airtight type) or from the last digester (if your plant has an open digestate storage tank or basin). Measure the VS of the digestate.

5. Copy the measured values in a table or a spreadsheet (date of sampling, dry matter (DM), VS of the feedstock or biomass mix (VS_{in}), VS of the digestate (VS_{out}), OL, and OLR).

6. Plot the data as historical graphs (VS_{in}, VS_{out}, OL, and OLR in the y axis; dates in the x axis). This will allow observing the plant's trends. Please note that the silage's VS tends to decrease with time because of its oxidation in the silo or trench, so it will be necessary to correct the total amount of feedstock, Q, so as to keep the OL and the energy production constant.

7. Should you notice that the VS_{in} decreases (for instance, because the silage absorbed moisture), then recalculate the quantity of feedstock, Q, to maintain the OL constant.

4.1.2 Measuring the Efficiency of the Feedstock's Conversion into Methane

The second practical application of the VS measure is checking the efficiency of the biomass' conversion into methane. This verification is useful for optimizing the purchase cost of the feedstock, if such is your case. Another application is controlling the regular operation of the biogas plant, because quite often the managers increase the OLR with the scope of stabilizing the biogas production, but the consequence of such action is reducing the net methane yield compared to the feedstock's BMP. *Increasing the OLR is like giving gas to a car's motor—the output power will increase, but the fuel consumption will increase more than proportionally.*

In biogas plants having an airtight digestate storage tank, the problem described earlier becomes marginal, because such tanks, being unheated, feature long SRT and hence serve as psychrophilic post-digesters. On the contrary, in biogas plants having an open storage tank, the incomplete digestion of the biomass caused by an increase in the OLR not only means wasting feedstock, but also emitting more greenhouse gases in the atmosphere. It is then the last type of biogas plant that requires a more frequent and accurate measure of the digestion efficiency. Figure 4.1, taken from a study performed in the United States in 1998, shows how the digestion efficiency decreases with increasing OLR (inversely proportional relationship).

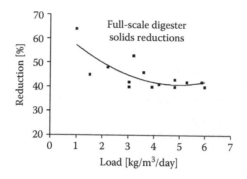

FIGURE 4.1
The efficiency of the feedstock's conversion into methane decreases with increasing the OLR of the biogas plant. (From Burke, *Dairy Waste Anaerobic Digestion Handbook*, Environmental Energy Company, Olympia, WA, 2001, who in turn quotes Lusk, 1998. With permission.)

If the procedure described in Section 4.1.1 has been correctly carried out, the biogas plant's manager will have all the elements to calculate the digestion efficiency. The latter is defined as:

$$\eta_{AD} = \frac{SV_{in} - SV_{out}}{SV_{in}}$$

The procedure to monitor the plant on the basis of the calculated η_{DA} is as follows:

1. Calculate η_{DA} with the said formula, using the data measured as explained in Section 4.1.1.
2. Save the information (date, measured value of η_{DA}) in a file containing the historical series.
3. If $\eta_{DA} > 0.45$, then the operation of the plant is acceptable. Lower η_{DA} values could indicate the existence of a hydraulic short circuit (if the trend is a constant reduction of the efficiency) or temporary partial inhibition of the process (in the case of a single value being lower than the average).

Table 4.1. Example of a spreadsheet, containing the necessary data to monitor the OL and the efficiency of the AD process. It is convenient to plot in a separate sheet the values of VS_{in}, VS_{out}, OLR, and η_{DA} as a function of time, so as to easily detect negative trends.

4.1.3 Corrective Actions in the Case $\eta_{DA} < 45\%$

It is important to note that η_{DA} seldom exceeds 60% in single-stage industrial biogas plants, because the microorganisms employ a fraction of the carbon

TABLE 4.1

Possible Template that the Reader Can Easily Prepare with a Spreadsheet

Week	DM_{in} [%]	DM_{out} [%]	VS_{in} [%]	VS_{out} [%]	OL	OLR	η_{DA} [%]	$\eta_{DA} > 0.45$?
n	15	10	11.8	6	30 ton·VS/ day	2.8 kg·VS/ [m³·day]	49	OK
$n+1$	13	11	11.5	6.4	30.1 ton·VS/ day	2.8 kg·VS/ [m³·day]	44	Check!

that constitutes the feedstock to build their own biomass (proteins, lipids accumulated in the cells) and to reproduce themselves. In CSRT-type plants, it can happen that, after 4 or 5 years of operation the efficiency η_{DA} tends to gradually, but constantly, decrease in spite of the OLR being kept within design limits. The causes can be two: either the partial inhibition of one of the steps of the AD process (usually the methanogenesis) or the reduction of the hydraulic retention time (HRT). It is possible to diagnose easily the first case with the biological tests explained in Section 4.2.4. Should no biological problem arise from the said tests, then the most probable cause of the decrease in efficiency is a *hydraulic short circuit*. We call that a reduction of the effective volume of the digester, caused by the accumulation of sediments in its bottom (as shown in Figure 4.2) and/or the formation of a mat of floating fibers on its surface.

Remember that, by definition:

$$SRT = \frac{V_{digester}}{Q_{in}}$$

Hence, if the input flow is kept constant so as to produce a given amount of energy, but the useful volume has become smaller because of the accumulated sediments, then the solids retention time (SRT) diminishes proportionally.

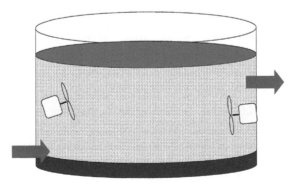

FIGURE 4.2

Reduction of the effective volume of the digester caused by sediment accumulation in its bottom.

Shorter SRT brings a consequence that the feedstock will be digested only partially, and the arousal of such condition is revealed by monitoring the VS and the digestion efficiency, η_{DA}.

Should the problem persist, the possible solutions are two:

1. Bring the digester's stirrers to their maximum power and closely monitor during the process. If the increased stirring is able to lift the sediments, then the digestion efficiency, η_{DA}, will begin to rise, until it reaches an acceptable limit.

2. If the former solution proves ineffective, it will be necessary to perform a test with chemical tracers, so as to check *if* the volume of the digester has been reduced and by *how much*. The said test is rather expensive and needs the assistance of an external chemical laboratory, both for defining the amount of tracer to load into the digester (at least 30 times the detection threshold of the available analytical instruments) and for performing the analysis of 30 samples. The general idea consists in loading the digester with a predefined quantity of an inert chemical product, which is not contained in the feedstock. Typically, lithium bromide (BrLi) or other salts of F, Br, and Li are employed.

   If the digester's stirring was perfect (and we assume this because the stirrers would have been running at full power during the process), then the concentration of BrLi will decrease in time according to an exponential law of the form:

$$C = C_0 \left[\frac{(V/q^2)}{T} \right] e^{-(t/T)}$$

where
   C_0 = initial concentration
   V = effective volume of the digester [m³]
   q = feedstock's volumetric flow [m³/day]
   T = nominal (or theoretical, according to the case) HRT (Remember: HRT = SRT because this test is valid only for CSRT digesters)
   t = day when the sample was taken (usually one sample a day, always at the same time, during the test's duration)

When monitoring daily the concentration C, the day when $C \approx 0$ (threshold limit of the analytical method employed) represents the *real* HRT of the digester. If a hydraulic short circuit effectively exists, then the measured HRT will be shorter than the design or nominal HRT of the plant.

Now we know the real value of the HRT, and since the daily feedstock's flow (assumed constant during the test duration) is known, then we can calculate the effective volume available in the digester (by just applying the definition of HRT). Having a clearer view of the problem's magnitude, the manager of the biogas plant can now decide with more certainty which of the following strategies to adopt:

1. If the biogas plant is still "young" (i.e., its commercial end of life is expected in many years in the future) then it may prove more convenient to stop the biogas plant, empty the digester, clean it, perform any other maintenance operation, and finally restart the plant.
2. Keep the stirring at maximum power for some months and check if the situation improves. It is a good practice to eliminate fibrous and high-solid materials from the digester's diet, so as to prevent the accumulation of indigestible matter, and to employ liquid biomasses (whey, sludge, glycerol, etc.) or easily degradable feedstock (oilseed cake, light food rests, nonedible flour).

Taking the correct decision is a difficult task, requiring thorough analysis case by case.

Practical Conclusions

By checking the digestion efficiency—at least monthly—and keeping a historical record, the biogas plant manager will be able to diagnose any problem as soon as it begins to manifest. If a slight but steady negative trend of the efficiency values becomes measurable, then the problem can be easily solved by increasing the stirring power and/or changing to a less "fibrous" diet. When hydraulic short circuits are diagnosed in time, there is a high probability of correcting the problem and avoiding the accumulation of sediments and floating mats, whose removal may eventually require stopping the plant and opening the digester to clean it.

4.2 Practical Applications of the BMP Test

Regardless of measuring unknown biomasses or reference substrates whose BMP is known *a priori*, the procedure to carry out this test is always the same, with some little variants already explained in Chapter 3. The following practical examples are useful in understanding the economic importance of the BMP assay.

4.2.1 The Right Price for the Feedstock

In many industrialized countries, investors build biogas plants with the only scope of producing energy, attracted by local policies that subsidize such kind of installations. If such is the reader's situation, then the following example will show how to optimize the feedstock's procurement activities.

A vendor offered one of the author's customers a "special feedstock with high biogas yield" at 150 €/ton. The customer's question was naturally: Is it worth?

The "special formula" in question, called here as "Substance X," was a dry fine powder, having an almost imperceptible smell. Its characterization gave the following values:

- DM = 90.94 % on w.w.
- VS = 85.04 % on w.w.
- Ash = 5.9 % on w.w.

The inoculum employed for the test was digestate sampled from the same plant, previously filtered to eliminate coarse fibers. Its VS concentration resulted to be 3.8%.

The assay was performed with an AMPTS-II, having real-time normalization of the measured methane volume, with its overestimation correction function activated. The tests were carried out in duplicate. The test temperature was set at 38.5°C (because the plant works in the range 38°C–39°C). The I/S ratio was chosen to be 3, resulting then in 4 g VS of Substance X to be loaded in each reactor. Table 4.2 shows the individual methane volumes

TABLE 4.2

Results of the Individual Reactors and their Averaged Specific Production

Day	Blank [Nml]	X1 [Nml]	X2 [Nml]	Net CH$_4$ Production [Nml/g SV]
0	0	0	0	0
1	554	934	1061	111
2	725	1161	1286	125
3	1028	1499	1620	133
4	1368	1987	2131	173
5	1600	2366	2580	218
6	1716	2628	2868	258
7	1812	2812	3063	281
8	1932	2959	3204	287
9	2079	3116	3347	288
10	2126	3285	3510	318
11	2150	3328	3588	327
12	2174	3359	3616	328
13	2197	3387	3642	329
14	2219	3414	3667	330
15	2240	3442	3691	332
16	2260	3471	3717	333
17	2277	3498	3738	335
18	2293	3522	3763	337
19	2309	3545	3784	339
20	2325	3572	3808	341
21	2338	3594	3828	343
22	2346	3608	3846	345
23	2346	3608	3846	345

produced by each reactor and the average result (5th column, showing the calculation of the specific production).

The net average methane yield is easier to analyze when plotted as a graph instead of as a table. Then the graphic function of the spreadsheet turns very useful, giving the result shown in Figure 4.3.

In general, the anaerobic degradability of such a substrate is good, because already in 12 days it yields 94% of its BMP. The curve's shape is somehow anomalous if compared to the classical sigmoid. The causes could be either that the inoculum is not "specific enough" to digest such substrate, or that Substance X is a mixture containing easily degradable sugars, complex polysaccharides, and or/proteins (more difficult to digest) and probably a small proportion of fats. The test's duration was just 23 days because "the result was required urgently." Such a short duration leaves a doubt about the *ultimate* BMP. According to the vendor, the measured BMP was lower than his own estimation. Not without reticence, the vendor finally admitted that Substance X contains "some fats." Fats have a high BMP. For sure, a small percentage of them would increase sensibly the BMP of the mixture, but on the other hand, their complete degradation requires 45–60 days. Then at least 60 days of testing would have been necessary to find out the "ultimate" BMP of Substance X. On the other hand, the customer's biogas plant is designed for an SRT = 30 days. So the BMP value that really matters for the customer is the yield in 30 days, and the test with 23 days is a good approximation, given the "hurry" to decide whether to purchase the feedstock or not.

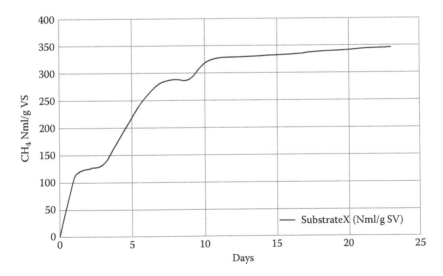

FIGURE 4.3
The complete anaerobic degradation curve of Substance X.

Now that we have all the data about Substance X's potential, calculating its economic convenience is relatively straightforward. Since the concentration of VS is 85%, then each ton of Substance X contains 850 kg, and will yield

$$Q_{CH_4} = 850 \ [\text{kg/t w.w.}] \times 0.350 \ \left[\text{Nm}^3/\text{kg}\right] = 297.5 \ [\text{Nm}^3/\text{t w.w.}]$$

The direct cost of the methane produced with such feedstock will be then:

$$C_d = 150 \ \text{€}/297.5 \ \text{Nm}^3 = 0.504 \ \text{€}/\text{Nm}^3$$

Considering that the generator has 38% electric efficiency,[*] and that the lower calorific value of methane is 9.94 kWh/m³, the direct cost of the electricity generation will be:

$$C_{el} = 0.504 \ \text{€}/\text{m}^3 \ /(0.38 \cdot 9.94 \ \text{kWh/m}^3) = 0.134 \ \text{€}/\text{kWh}$$

The said cost is then the "fuel cost" of the plant, supposing to feed it with Substance X. Adding the costs of mortgage, labor, lubricants, maintenance, etc., will result in the total cost per kWh, which must be then compared to the feed-in tariff, so as to decide if the substrate in question is economically convenient or not.

In this special case, the BMP declared by the vendor made the substrate to appear slightly positive with the feed-in tariff of the customer, but the effective BMP measured in the lab, in the assumption of 30 days HRT or shorter (real operational condition of the plant) was too low to justify the price required by the vendor.

4.2.2 Optimizing the Biogas Plant's Diet

Like any other living being, microorganisms will grow healthier if they are fed with an assorted and balanced diet. Furthermore, if the biodiversity of the entire bacterial ecosystem is rich, its digestion performance will be better. The following rules will help the biogas plant manager in the practical optimization of the digester's diet:

1. Mixtures of different substrates will generally work better than "pure" substrates (only carbohydrates, or only protein, or only fats).
2. 2 + 2 will not always yield 4 in the AD reign. A mixture of two substrates will yield its maximum methane potential if their respective

---

[*] Author's note to the English edition: Biogas plants in Italy usually sell only electricity. Some amount of heat is employed to keep the digester's temperature, and the excess heat sometimes has marginal uses for greenhouse heating. In our example, excess heat is just dissipated to the atmosphere, hence in this calculation the economic value of heat is null.

proportions are such that the C/N/P ratio results as near as possible to 150/5/1.

3. Mixtures of manure from different animal species, having different bacterial floras, will be more biodiverse than the pure manure of a single species. In general, such mixtures will be able to degrade a wider range of feedstock.

4. Ruminant's manure (cows, buffalos, goats, sheep, camels, reindeers, etc.) is a suitable inoculum for digesting cellulose-rich substrates. Monogastric animals' manure (pigs, chickens, ducks, humans, etc.) is a suitable inoculum for digesting protein and fat-rich substrates. Manure from nonruminant herbivores (rabbits, horses, elephants, etc.) usually has a higher BMP than that of ruminants, because such animals have a different bacterial flora allowing them to digest cellulose and hemicellulose (Zhao et al., 2012), but are less efficient than cows in chewing their fodder, since they only chew once. Hence, the said manure is a good substrate and an acceptable inoculum to start a biogas plant too.

Figure 4.4 shows an example of optimization performed empirically. A biogas plant running regularly on cow manure had the opportunity to incorporate rabbit manure. The questions of the biogas plant's manager were "Is the addition of rabbit manure beneficial?"; "Will its high concentration of ammonia upset the process?"; and "Which is the best proportion to mix both substrates?" The tests were performed in the author's lab, using an AMPTS II and digestate samples from the customer's plant as inoculum. Cow and

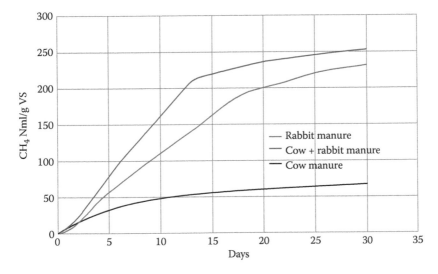

FIGURE 4.4
Example of the results that can be obtained by just mixing substrates in the right proportions.

rabbit manures were mixed in different proportions, and all were digested at the same temperature (38°C, so as to favor the maximum biodiversity). The figure shows how the specific production of one of the mixtures yielded much more methane than the single substrates considered individually.

Such a simple test gave as a result the ideal mixture proportions to maximize the methane yield.

4.2.3 Preventing the Inhibition Caused by "Difficult" Substrates

In this context, a "difficult" substrate is any degradable substance that, in spite of its theoretical degradability, can upset the AD process because it contains some inhibiting compound. Some examples of such substrates are chicken dung (high ammonia and hydrogen sulfide), vegetable oil extraction byproducts (containing long-chain fatty acids and waxes), olive mill wastewater (containing polyphenols), and cheese whey and marine biomass (containing salt). Sometimes the concentration of inhibiting substances in the substrate would not pose any problem for a "healthy" inoculum, but in some biogas plants (especially in thermophilic ones), the bacterial ecosystem has little biodiversity and hence can suffer a biological collapse if the diet is suddenly changed.

The following is an extreme example that shows how a degradable substrate having a complex chemical composition can upset a biogas plant that "worked well" (but at the edge of the biological collapse, without the manager being aware). The biogas plant in question runs on only corn silage, without the addition of any kind of manure. The fermentation was induced and maintained in time by the regular addition of "a special formula covered by industrial secrecy" (presumably a mix of lyophilized bacteria and probiotic substances). The plant's owner pretended to replace part of the silage with slaughterhouse waste, composed mainly of lipids and proteins. Figure 4.5 shows very eloquently the disaster that the said sudden change in the digester's diet would have caused if no laboratory test had been performed in advance.

When sampling the inoculum, the plant manager reported: "It's all in order, because the FOS/TAC value is perfect." We can deduce from the curve of the reactor *Acetate* (dashed line) that the addition of acetate caused a partial inhibition of the inoculum, a fact demonstrated by 10 h of lag phase. It is possible to deduce from this test's results that the inoculum already contained a certain quantity of propionic and butyric acids, which the test FOS/TAC, performed by the biogas plant manager, measured as acetic acid. Furthermore, we must recall that this inoculum had a low buffer capacity, because the plant did never receive manure, and was not degassed before starting the test, with the scope of simulating what would happen in the case the slaughterhouse waste was added to the digester. Adding a lipoprotein-based substrate produced immediately a remarkable methane amount during the first 10 h (probably because it created a favorable C/N ratio, or because the said

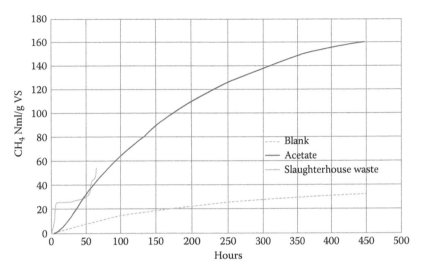

FIGURE 4.5
Example of inhibition caused by a "difficult" substrate because the inoculum's bacterial biodiversity and overall biological activity was scarce.

substrate contained a small fraction of easily degradable amino acids). The surge of methane production was then followed by a short *plateau* (beginning of the inhibition) and subsequently a short burst, suddenly interrupted (biological collapse).

4.2.4 Checking the Digestion Efficiency of the Biogas Plant

We have seen in Section 4.1.2 how to check the efficiency of the biogas plant in converting biomass into methane. Furthermore, it is important to evaluate the residual methanogenic potential of the digestate, to quantify its residual energetic value and eventually to analyze possible strategies to extract the maximum profit of it.

The following example was taken from a real case, a biogas plant having one single digester and an open tank for collecting the digestate. The digester was fed with a mixture of swine and bovine manure, corn, and triticale silages.

The VS concentration of the feedstock mixture sampled from the compensation tank placed before the digester was 8.75%, while that of the digestate, sampled at the digester's output, was 4.1%. The conversion efficiency in methane, according to the VS reduction criterion, was hence:

$$\eta_{DA} = \frac{8.75\% - 4.1\%}{8.75\%} = 53\% > 45\%$$

The performance of the plant is good, although not at all brilliant. However, the formula of the VS reduction efficiency does not allow us to deduce if the residual fraction of the VS contained in the digestate is still capable of producing useful amounts of methane. To recover such residual methane, the simplest strategy possible in CSRT plants is just to increase the HRT. Hence, the digestate storage tank should be covered with an airtight dome and if possible heated too, so as to extend the digestion until the complete depletion of any degradable organic matter. For the plant owner, the important question was: "Is it profitable to invest money in covering the open digestate storage tank with a biogas collection dome?"

The biological test in such cases is very simple: it is enough to incubate the same quantity of sludge, sampled from the premix tank before the digester and from the digester's output, and calculate the fraction of residual BMP in the latter. Please note that a consistent percentage of residual BMP in the digestate, e.g., 20% of the feedstock's one, not only means losing 20% profit, but it is also a source of greenhouse gas emissions to the atmosphere. Figure 4.6 shows the result of the test.

It is quite evident from the curves that in this case the residual methanogenic potential of the digestate is still high, 50 Nml/g VS, while the feedstock mixture has 103 Nml/g VS. The resulting efficiency is hence 51%, a value coherent with the one estimated with the VS reduction criterion. Then, it is worth performing an economical evaluation of the investment in a biogas collection dome for covering the digestate collection tank. Such investment will allow saving nearly 50% of the cost of silage for the rest of the plant's commercial life. Another option is to not cover the digestate tank and to replace the current silage with alternative substrates having a shorter digestion time (i.e., less cellulose). In such cases, there will be no investment cost, but the running cost

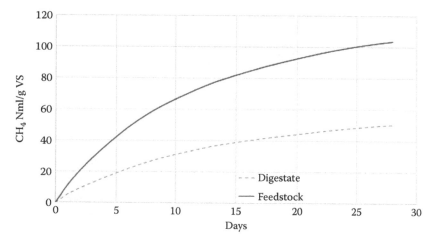

FIGURE 4.6
Residual methanogenic potential of the digestate, compared to that of the feedstock.

will be higher, since easily degradable biomasses (e.g., molasses, sugar beet pulp, alfalfa, etc.) are usually more expensive than corn and triticale silage.

4.2.5 Determination of the Optimum SRT/HRT

Most of the biogas plants in Europe are designed to run with SRT ranging from 30 to 60 days, necessary for the complete anaerobic degradation of the most common substrates. In some cases, alternative substrates having short digestion times may be available (e.g., byproducts from the food industry like potato skins, wastewater from beverage production, etc.). Figure 4.7 shows the digestion curves of wastewater from a beverage factory. The organic matter content of the said industrial effluent is very high: 69,600 mg/l of COD. It is composed mostly of residual starch and dissolved sugars, so their anaerobic degradation is quick, as clearly shown in Figure 4.7.

We can observe that the curve's shape has some "hump," but anyway it is almost linear until the complete depletion of the digestible fraction of organic matter, on the 12th day. In such a case, we can employ the liquid to dilute silage or any other solid feedstock composing the usual diet of the biogas plant, reducing proportionally their quantity. Caution must be taken to prevent the acidification of the digester. The conversion of sugars and starch into acetic, propionic, and butyric acids is faster than the conversion of said acids into methane. If the inoculum has enough alkalinity, or if the manure is added as cosubstrate, there is little risk of acidification. But in industrial plants that do not employ a high alkaline cosubstrate like manure, it may be necessary to add alkalinity, usually as sodium carbonate or bicarbonate, or lime, to neutralize the pH. In our example, the wastewater had pH = 6.4 when

FIGURE 4.7
Anaerobic degradation of wastewater from a beverage industry.

starting the test. Since the inoculum was sludge from an agricultural biogas plant running on cow manure, neutralizing the substrate before starting the test was not necessary.

4.2.6 Determining the Efficacy of Additives and Pretreatments

The use of commercial additives is probably one of the most discussed (and discussible) arguments when dealing with the optimization of a biogas plant. The author's personal experience on this subject is that most of the arguments in favor of the said praxis are self-referenced pitches of the vendors or some user's belief that his plant will not work without the addition of probiotics, trace elements, desulfurants, etc. The arguments of the vendors are usually weak and range from void phrases such as "thousands of plants in Germany use our product" to "scientific studies made by the University of XY show that this product boosts the biogas production." Of course, when checking the source one finds that the "scientific study" was sponsored by the vendor and in some cases the result shows a small improvement of the total biogas production, without indication of the total error margin of the test or the net production of methane. Some examples of said situations will be analyzed in Chapter 6. The factual reality is that the efficacy of a given additive varies from one biogas plant to the other, or even at different times in the same plant, depending on the initial state of the microbial consortium. In some rare cases, it is possible to observe miraculous performances, in most cases nothing happens, and in some special cases, the additive may even induce a slight inhibition. Figure 4.8 shows the effect of a product, consisting on live lyophilized bacteria, on an inoculum sampled from a biogas plant

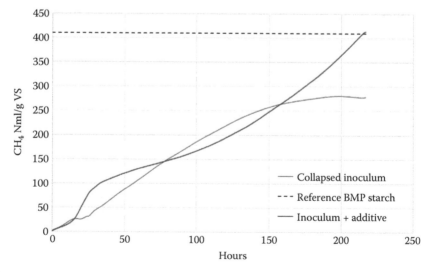

FIGURE 4.8
The effect of adding live bacteria to an inhibited inoculum.

that had suffered a biological collapse. The reference substrate employed for the test was starch. It can be observed that the reactors to which the additive was added reached the theoretical BMP of starch in 10 days, while the control reactors, loaded with the same amount of starch but having the untreated inoculum, were not able to degrade the said simple substrate.

The curve corresponding to the inoculum treated with the additive (red line) appears somehow "anomalous" (very different from the usual sigmoid). The cause is the sudden change in the composition and specific biological activity of the microbial ecosystem. Adding a new bacterial population to an inoculum that already had some sort of equilibrium caused a series of variations in the metabolic rate of the system, until the microbial ecosystem reached a new state of equilibrium after 100 h. From that moment onward, the methane production became constant (almost linear growth). The control reactors instead (continuous line) had their bacterial ecosystem already in equilibrium, although the biodiversity was poor. The result is a curve that is indeed a sigmoid, but very "flat" and having a lower asymptote than the starch's BMP. This means that the inoculum was not capable of degrading all the starch, even after 10 days.

Figure 4.9 shows the result of the digestion of shredded mixed waste from a pig slaughterhouse, employing a "healthy" inoculum. A comparative test was run, adding the same product of the former example to this inoculum and also boiling the waste but without adding any product to the inoculum. It is quite evident that boiling lipoprotein-based substrates like the one in the example increases remarkably their BMP, while the difference between the digestion of the raw waste, with and without additive, is almost irrelevant (the difference between both curves lies within the uncertainty range of the test).

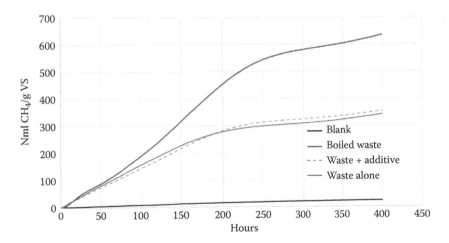

FIGURE 4.9
Comparison between the influence of a thermal pretreatment and the use of a commercial additive in the BMP of swine slaughterhouse waste.

Another comparative test, this time between additive *A* (defined as "probiotic mixture of alkaloids and trace elements"), additive *B* (defined as "mixture of lyophilized bacteria, vitamins, and biocatalysts"), and the untreated inoculum gave the results shown in Figure 4.10.

Observing the curves, it is evident that *additive A* promotes the bacterial activity in an almost "miraculous" way. Nevertheless, the test showed that the concentration of methane diminished remarkably (below 50%!). This behavior is a clear demonstration that the additive has a positive effect on the fermentative bacteria, but little or no effect on the methanogenic flora. Furthermore, the curve of additive *A* shows "humps," while those corresponding to the blank and to additive *B* are almost perfect sigmoids, but "flat." Hence it is necessary to thoroughly evaluate if, from the operational point of view, the use of additive *A* is worth. Figure 4.11 shows the net methane flow of the same reactors, i.e., the time derivative of the curves shown in Figure 4.10. In this case, the said curves were obtained automatically with an AMPTS II, because it is a built-in function of the said instrument, but it would have been anyway easy to obtain them with a spreadsheet and daily readings of a manual instrument. Please note how *additive A* induces an irregular bacterial activity, showing remarkable production peaks followed by valleys (the "humps" in the curve in Figure 4.10, which may seem almost imperceptible to the unexperienced eye). Such an erratic production of gas is a drawback of *additive A*, because the flow peaks may lead to waste biogas in the plant's torch, while during the flow valleys the amount of methane could be insufficient to feed the generator at nominal power.

In the case just analyzed, the batch test provides useful information to understand the benefits and drawbacks of additive *A*, but it is not enough to

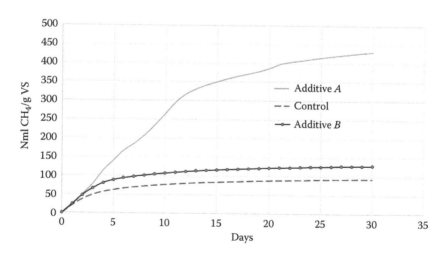

FIGURE 4.10

Comparison between the net methane yield of the untreated inoculum (control) and the same inoculum with two different commercial additives.

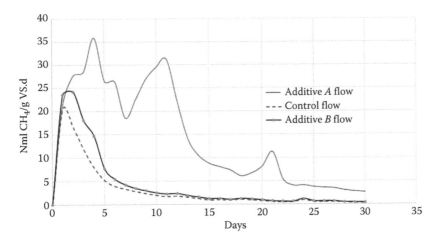

FIGURE 4.11
Net daily flow of methane produced by the control reactor and with two different additives.

define an operational strategy. The author's customer had requested to perform the test on the unfiltered inoculum, sampled from the digester, with and without additives. Such procedure is not reliable, because the peaks observed in the batch test could be just the result of a transient situation, for instance, undigested organic matter accumulated in the inoculum that the additive made digestible. In such cases, it is quite advisable to carry out a simulation of the process in a continuous reactor, at different doses of additive *A*, and even the simultaneous application of both additive *A* and *B*, so as to check if a stationary methane production can be achieved. Nevertheless, continuous tests are more time and labor consuming than batch ones. A cheaper alternative solution is to perform some comparative batch tests with different reference substrates. The procedure should be the regular one—three samples of filtered inoculum should be treated with the additive, and three other samples left untreated. All six samples should be degassed before performing the test with the chosen reference substrate. Such standard procedure takes longer than the one chosen for the experiment presented earlier, but leaves no space for doubts when analyzing the results.

Practical Conclusion
Employing commercial additives in a biogas plant does not necessarily improve the process. It is very advisable always to measure in the lab the effective need, the necessary dose, and the eventual side effects (e.g., an increase in total biogas production accompanied by a reduction of the methane percentage). Once the objective data under controlled conditions are obtained, it is easy to plan a rational strategy to boost the real-scale process or to refrain from spending money in useless products. The same criterion applies to pretreatments. It is necessary to check if they induce a beneficial or an inhibitory effect, and if the application costs (both economic and

thermodynamic, i.e., considering each electric kWh consumed in the pre-treatment as roughly 0.27 Nm³ of methane) are worth the eventual benefits. Some additives contain heavy metals, for sure in very small quantities, but if the digestate is employed as fertilizer, then said pollutants end up in the soil and are absorbed by the crops, or may be leached to the underground waters when it rains. This environmental aspect must be carefully evaluated, without any prejudice, but on the basis of objective figures.

4.2.7 Frequent Errors in Planning and Performing Biological Tests

Sometimes it may happen that the result of a biological test causes some perplexity, either because too different from the expectations or because the specific methane production curves do not have the classical sigmoid shape. A very frequent error is not degassing the inoculum enough before starting the test. In such cases, the "background noise" induced by the inoculum, combined with some "difficult" substrate (e.g., containing large amounts of lipids or any inhibitor) will deform the curves of specific methane production. Figures 4.12 and 4.13 show examples of incorrectly performed tests. The inoculum was sampled from the digester, and employed "as it was," without filtering and degassing it. The substrate to be tested contained a certain quantity of fats. Figure 4.12 shows how the curves of raw methane production look "quite normal," but after 30 days of incubation they have not reached the plateau, which means that the substrate's digestion was only partial. Such behavior was expected, since lipids usually need 60 days for their complete degradation. What the author did not foresee was the inoculum's behavior. In Figure 4.12, one can observe that between the 16th and the 21st day the curve has a "hump." Probably, the cause of such anomalous

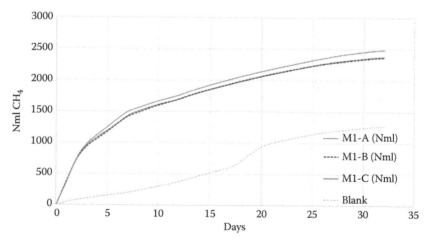

FIGURE 4.12
Individual CH₄ production curves of the blank and the sample reactors.

behavior was some amount of organic matter that was still in the hydrolysis and acidogenesis phase when the inoculum was sampled from the reactor.

This example was featured by some partial inhibition caused by the fats, and by employing an inoculum without degassing it in advance (business as usual—the customer wanted quick results!). As it was to expect, when plotting the net CH_4 production curves (i.e., average of the samples' production minus blank's production, divided by the VS), the result showed the anomalous shape that can be observed in Figure 4.13.

When the BMP test is performed correctly, i.e., thoroughly degassing the inoculum in advance, the specific production curve must be always *increasing* because, by definition, the said curve represents the cumulated gas volume and hence cannot have a decreasing inflection as the one shown in Figure 4.13. In this case, two factors contributed to yield a very anomalous specific production curve: having accepted the customer's request of employing a "fresh" inoculum (hence a high "background noise") and the partial inhibition caused by the fats. Both factors became evident from the 17th day, when the daily gas flow produced by the blank became higher than in the former days, while the sample reactors had the opposite behavior, resulting in a decreasing net production curve. If the customer had not been in a hurry and the inoculum had been correctly degassed in advance, the curve would have been a perfect sigmoid, most probably showing a plateau from the 50th day, when the fats usually reach their complete degradation. In the present particular case, continuing the test up to 60 days—as established by the VDI 4630—had no practical sense, because the biogas plant in question was designed for a nominal SRT of just 30 days, and furthermore the digestate collection tank was of the open type. In such particular cases, the plant manager must always remember that it is useless to buy a feedstock with a higher

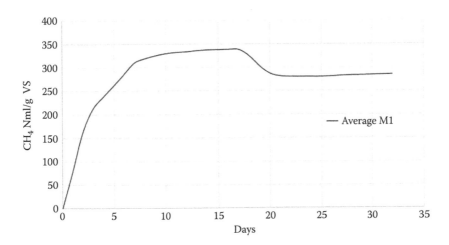

FIGURE 4.13
Example of an anomalous specific production curve.

BMP (as all substrates containing lipids and proteins are) if the plant's SRT is shorter than the total time necessary for the total degradation of such substrate. In this very particular example, the biogas plant manager had the possibility of getting such substrate for a very good price, if purchasing it before a certain offer expiry date, which was the reason for the hurry in performing the test and not degassing the inoculum. Consequently, the practical calculation of the digester's diet was based on the BMP measured on the 15th day.

Important
The rules for planning and performing AD tests described in Section 3.8.1 are not just "academicians' fastidiousness." Not following the given procedures means taking risk to obtain useless results, and wasting time and money in repeating the test. Remember, performing anaerobic digestion tests, even in the plant's laboratory, is easy and does not require a degree in Chemistry. All you need is just common sense, paying attention while preparing the test, and following the procedure without exceptions.

4.3 Using Reference Substrates to Check the Hydrolytic Activity: How to Find Out if Something Is Going Wrong

Together with the specific methanogenic activity (SMA) test, which we will study in detail in Section 4.4., it is advisable to perform the hydrolytic activity tests at least once a month, so as to assess if the bacterial ecosystem is healthy and to prevent the plant's biological collapse. We have already introduced the BMP of the reference substrates in Chapter 3, Table 3.3., and in Sections 3.10.1. through 3.10.5. We have seen the special procedures and the peculiarities of some of the said substrates, requiring I/S ratios higher than 3 to prevent the inhibition caused by the volatile fatty acids (VFA), by the presence of Na ions or long chain fatty acids (LCFA). In the next example, taken from a real-life situation, we will learn how to interpret the test's results.

4.3.1 Hydrolysis Test of Cellulose

Cellulose-degrading bacteria thrive exclusively in the digesting tube of ruminants, concretely in their rumen. Other animals, including people, are incapable of digesting cellulose, a fact employed for some extremely dangerous slimming diets, since eating toilet paper or cellulose in powder fills the stomach and provides a satiety sense but does not contribute with calories. If during the hydrolytic activity test of cellulose, the inoculum is not capable of reaching the reference BMP within 20–30 days, the biogas plant manager should immediately proceed to inoculate the digester (or the pre-fermenter, if the plant has two or more reactors) with cow, sheep, or goat manure. The inoculation will be even more effective if rumen can be obtained from a slaughterhouse. Some commercial probiotic products contain lyophilized bacteria, but their cost is usually high, so it is necessary to check their efficacy in the laboratory before deciding their purchase. If the hydrolytic activity test fails or shows a very slow degradation rate, it is advisable to repeat it with different doses of rumen, commercial additives, manure, or even digestate from another biogas plant, so as to determine the minimum amount of additional inoculum necessary for reestablishing an adequate bacterial population.

4.3.2 Hydrolysis Test of Sugar and Starch

This test can give false-negative results if the inoculum has little *buffer* capacity, or if it has already some amount of accumulated VFA, caused by other biological problems. Sugar and starch are simple, easily degradable carbohydrates. Their degradation produces alcohols, which in turn are further fermented into VFA. If the hydrolysis rate is higher than the rate of conversion of acetic acid into methane, then the reactor will tend to acidify. Hence, this test is complementary to the SMA test and much more useful than the FOS/TAC (VFA/TA) test, providing a more complete overview of the eventual unbalances of the sludge's bacterial ecosystem. It is customary in some countries to add alkalinity to the inoculum, so as to neutralize the acidification caused by starchy or saccharine substrates (e.g., by adding sodium bicarbonate to the digester). Such practice only masks the symptoms without correcting the causes, and must be avoided for two simple reasons: sodium bicarbonate costs money, and the presence of additional Na ions can inhibit the process. The said argument will be analyzed in detail in Chapter 6. It is preferable to check the effective hydrolytic activity in the laboratory and, if necessary, to add manure to the digester. Adding manure not only costs nothing in most cases, but it also provides enough alkalinity and additional living hydrolytic bacteria.

Practical Example

Figure 4.14 shows the anaerobic digestion curves of three products available in any supermarket, which were employed as reference substrates for testing the hydrolytic capacity of an inoculum, sampled from a biogas plant that was not operating satisfactorily. It can be observed that their BMP is not much different from the theoretical BMP of the analogous pure laboratory reagents. Toilet paper tends to yield BMP values near to the minimum because it is not 100% cellulose. In this sense, natural cotton in flocks is better than toilet paper, since it is composed of 99% cellulose (Remark: Avoid cotton disks because they usually contain some wax). The shape of the anaerobic degradation curves of both starch and sugar does not follow the classical sigmoid pattern, but looks rather anomalous. It is impossible to establish *a priori* if such anomalous shape is the result of the acidification caused by the high degradability of these substrates combined with a low buffer capacity of the

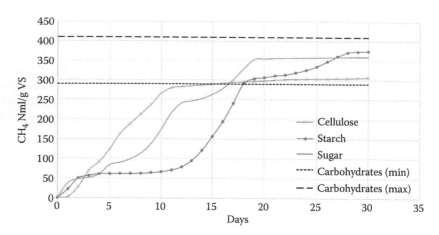

FIGURE 4.14
Anaerobic degradation curves of toilet paper (cellulose), white sugar (sucrose), and corn starch.

inoculum, or if the latter has a low hydrolytic capacity, or if the real problem is a scarce methanogenic activity. The last supposition seems to be the most probable in this case because the degradation curve of cellulose (a more complex molecule, hence more difficult to digest) looks almost like a perfect sigmoid, although somehow "flat." All three reference substrates yielded an acceptable BMP, so it seems that the hydrolytic capacity of the inoculum can be considered "normal." To understand why the curves show such strange shapes, it is then necessary to compare the result of Figure 4.14 with the results of the additional tests that will be explained in the next paragraph.

4.3.3 Hydrolysis of Proteins

The hydrolysis of proteins is always slower than that of simple carbohydrates, but if the inoculum is "healthy" the complete degradation of protein should take about 10 days. The eventual preexisting accumulation of ammonia in the inoculum, plus the ammonia produced by the N contained in the proteins (typically 8% of the VS) can sometimes trigger inhibition phenomena. When encountering any problem with the proteolytic capacity test, it is necessary to check if the cause is the inhibition by ammonia excess. A second test must be carried out, loading the reactor with the same total quantity of VS (same I/S = 3 as the proteolytic test), but in this case half of the VS will be protein and the other half will be cellulose. If the production of methane results in nearly the average of both substrates' of BMP, then the eventual problem is the accumulation of ammonia and not a lack of proteolytic capacity. If a spectrophotometer is available, measuring the ammonia content of the inoculum will turn useful, since concentrations above 3000 mg/l of ammonia are usually considered an alert limit by some authors. Nevertheless, the process temperature must also be considered, since the inhibition caused by ammonia is more probable in thermophilic plants, and not so troublesome in mesophilic ones. Should the ammonia be the cause of the low proteolytic activity, the solution consists in modifying the diet of the digester, reducing the percentage of protein- or ammonia-rich substrates (e.g., swine and chicken manure) and increasing the percentage of vegetal matter, which is usually richer in carbon than in nitrogen.

Should the ammonia levels be acceptable, then the scarce proteolytic activity can be improved by adding manure of monogastric animals (chickens, pigs, humans), which contain traces of proteolytic enzymes and a bacterial flora more suitable to hydrolyze proteins compared to that of ruminants or herbivores in general.

Practical Example

Figure 4.15 shows the results of the hydrolytic activity test using fish gelatin as reference substrate, with the same inoculum employed for the test shown in Figure 4.14. Once again, it can be observed that the inoculum, even reaching the minimum acceptable reference BMP, presents scarce specificity to digest such substrate. The shape of the curve is more regular compared to the curves produced by the digestion of sugar and starch. We cannot state *a priori* whether the bottleneck of the process is the proteolysis of the substrate or the final conversion step to methane. In this particular case, all points to the second supposition, as already stated in Section 4.3.2—the inoculum has a scarce SMA.

4.3.4 Lipolytic Activity Test

This test is performed seldom, but it must be considered mandatory if planning to feed the biogas plant with fatty biomasses (e.g., olive mill pomace,

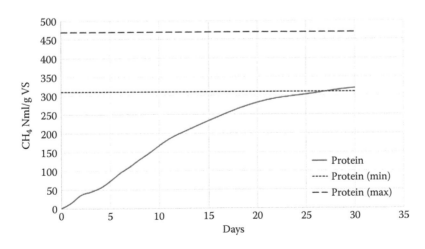

FIGURE 4.15
Proteolytic activity test using fish gelatin from the supermarket as reference substrate.

slaughterhouse waste, oilseed cake, kitchen waste, etc.). Always remember that if the plant is designed for HRT <60 days, attempting to feed it with fatty substrates will result in a failure, since the feedstock will not have enough time to fully degrade, and furthermore, the presence of fats may hamper the full degradation of other substances composing the diet. Should this test give negative results, the possible solutions are:

1. To inoculate the digester with swine manure.
2. To employ commercial additives containing lipase, which is the enzyme that hydrolyzes lipids, usually extracted from swine pancreas and from the bile of any animal. In this case, the necessary dose may reach up to 1 g/l of pure enzyme so as to obtain a quick hydrolysis (4 h), resulting in a high operational cost. It is then necessary to perform laboratory tests, with different doses of enzyme, then check the minimum dose that allows full degradation of the substrate within 15 or 20 days HRT, and finally to evaluate the operational cost and overall economical convenience.
3. To inoculate the digester with anaerobic sludge from a sewage treatment plant, if the local norms allow it.[*]

Alternatively, the slowest but the cheapest and surest way to employ fatty feedstock in a biogas plant consists in taking profit of the Darwinian selection. Perform the lipolytic activity test with several I/S ratios, find out the minimum one that does not inhibit the process, and start feeding the plant with such maximum dose.

Example

Suppose the digester has 1000 m³ and its unfiltered sludge is 7% VS. After sieving the sludge, it has 3.5% VS. Suppose that the fatty feedstock is slaughterhouse waste containing 40% VS. Test the fatty feedstock with I/S = 3, I/S = 5, and I/S = 10. Suppose the tests show that I/S = 10 does not inhibit the process (perfect sigmoid shape, perhaps a bit flat). Since the amount of "live" sludge in the digester is:

$$I = 0.035 \cdot 1000 \text{ tons} = 35 \text{ tons VS}$$

---

[*] Note of the author to the English edition: in Italy and some other European countries, using sewage sludge in an agricultural biogas plant may represent a legal problem, because it is forbidden to employ the resulting digestate as fertilizer. From a scientific point of view, such regulations are absurd and even contradictory with the EU Directives on circular economy and "end of waste status." On the opposite side, the Swedish regulation *SPCR 120—Certification rules for digestate from biowaste by the quality assurance system of Swedish Waste Management (December 2007)* is probably the most pragmatic and scientifically rigorous example of digestate management policy in Europe. According to the said regulation, the acceptability of the digestate for agriculture depends just on its quality, evaluated on the basis of well-defined chemical analysis. Anyway, the biogas plant manager must be aware of the applicable local regulations before taking any decision.

The maximum daily dose of fatty feedstock should not exceed 3.5 tons VS/day, which means 8.75 tons of fresh slaughterhouse waste. As a precautionary measure, start feeding your plant with less than that, for instance, 7 tons/day.

N.B.: Fatty feedstocks usually have a long lag phase, hence it is probable that during the first days the biogas production drops. To avoid loss of production, it is possible to add some easily degradable feedstock during the first week, or start feeding a fraction of the maximum fresh feedstock dose and increase it gradually in the subsequent days.

After 1 month, the bacterial consortium will have adapted to the new diet containing fats, so the test should be repeated, so as to measure the new lipolytic activity and new maximum admissible dose. After a few months of gradual adaptation, the plant should be able to digest consistent proportions of fats in its diet.

4.3.5 General Substrate Inhibition Test

Sometimes it may happen that a batch of waste or byproduct from the food industry, or any other agricultural waste, is available at an attractive price for the biogas plant manager, potentially replacing the usual silage or commercial feedstock. The spontaneous question that the plant manager will pose to himself/herself is: "Will this stuff work? Won't it block the process?" Such questions are very pertinent and one should always be aware that, if a batch of agroindustrial feedstock is very cheap or even free, the most frequent reason is that it is not apt for animal consumption or other uses. If a substance is toxic for an animal, it may present some toxicity for microbes too. Frequent examples are straw that caught rain and got mold, meals and extracts from the oil industry treated with mineral acids or other chemicals, slaughterhouse waste or byproducts not suitable for human consumption, or pet food production. In other cases, the chemical features and consequently the BMP, together with the product's seasonality, make impossible to define a stable market price (the typical case of olive pomace in the Mediterranean area). All mentioned byproducts are examples of potentially (but not necessarily) inhibiting substrates that can upset the anaerobic digestion process. In general, the anaerobic bacteria manage to digest even toxic substrates, but that may require longer times and lower concentrations than the usual feedstock, so as to avoid triggering a biological collapse. In such cases it is a good practice to measure the BMP, degassing the inoculum during at least a whole week, replicating the test with different I/S ratios, as explained in the former example. We can assume I/S = 10 as a practical limit, since a substrate capable of upsetting the process, even when fed in small amounts, poses more risks than benefits. The following example, taken from real life, shows the risks of adopting residual glycerol from the biodiesel industry as substrate for AD. The "literature" and "BMP databases" employed by biogas plant manufacturers and engineers give a misleading idea of said substrate, since its BMP is very high, its density is higher than that of water, and being liquid facilitates

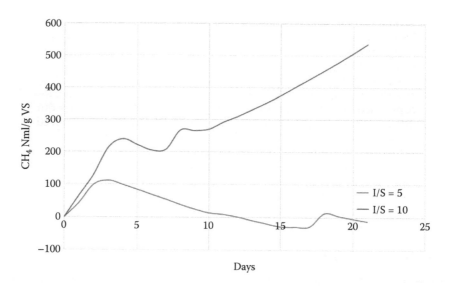

FIGURE 4.16
Inhibitory effect of residual glycerol on the AD process in a plant that has never employed such a substrate.

its handling. It may seem from such features that residual glycerol is the best AD feedstock one could imagine. What the "literature" does not say is that residual glycerol usually contains relevant quantities of soap, unsaponifiables, sodium chloride, or sodium sulfate, all of these substances with a certain inhibiting power for the archaea. Furthermore, the first step of the hydrolysis of glycerol is propionic acid, which we already know is a strong inhibitor of the methanogenic archaea. Figure 4.16 shows very eloquently the inhibiting effects of residual glycerol on the inoculum of a biogas plant running usually on corn silage and cow manure.

4.4 Applications of the SMA Test: Preventing the Biological Collapse and Selecting the Best Inoculum to Start a Biogas Plant

The SMA test should be performed at least once a month, instead of the FOS/TAC (VFA/TA) test, since the latter test has scarce utility to prevent biological collapses. It is useful to remember that the FOS/TAC test is not selective, because it measures the mix of VFA present in the sludge *as if* they were pure acetic acid. The simplified SMA test proposed here (employing only acetic acid, or vinegar, or sodium acetate) is easy to perform, requires only from 3 to 5 days to obtain a reliable result, and provides very useful information on the bacterial ecosystem's health. Should propionic and butyric acids have accumulated in the sludge, as a result of some unbalance in the microbial population, adding acetic acid to it will provoke a partial inhibition of the acetoclastic archaea, resulting in an SMA value smaller than a given acceptable standard. Remember that the information obtained with the AD test of the acetic acid (or sodium acetate, or just common vinegar) provides two useful control parameters: the SMA (quantitative) and the methanogenic capacity (qualitative).

4.4.1 Practical Example on How to Test an Inoculum Suspected of Methanogenic Inhibition Using Wine Vinegar

Figure 4.17 shows the result of a methanogenic capacity test performed with white wine vinegar as reference substrate, without neutralizing it. We can observe a perfect correspondence with the theoretical BMP of a 6% acetic acid solution. This test was performed with I/S = 5. The inoculum was the same that was employed for the tests described in Figures 4.14 and 4.15. Comparing the different tests (sugar, starch, protein, and wine vinegar) from a superficial analysis, it may seem evident that the inoculum under test has indeed methanogenic capacity, its proteolytic capacity being rather mediocre, but anyway acceptable. One could be tempted to assume that the hydrolytic activity is somehow compromised, since the curve in Figure 4.14. shows an unsteady behavior. Considering that the degradation of both protein and cellulose show steadier dynamics, and that said substrates are more complex to digest than simple sugar and starch, we can deduce that the anomalous

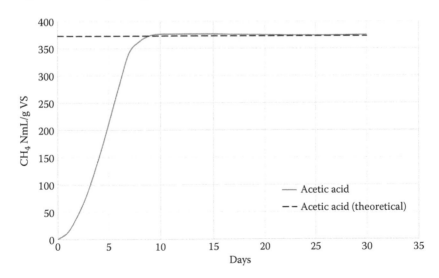

FIGURE 4.17
AD of white wine vinegar, employed as reference substrate, compared to the theoretical BMP of a 6% acetic acid solution.

shape of the methane production curves of sugar and starch is the result of the hydrolysis speed being higher than the methanogenesis speed. This practical case shows then that checking just the methanogenic capacity of the inoculum is not enough. The methanogenic capacity is a qualitative test: It tells whether acetoclastic methanogenic archaea are present or not, by simply checking if acetic acid (or sodium acetate) produces methane and is completely degraded. The test does not say *how active or numerous* the acetoclastic Archaea are. Thus, it is more useful to check the SMA, as will be described in the next paragraphs, since the figure gives *an objective measure* of the archaea's population consistency and activity.

Figure 4.18 shows the result of the SMA test. It employs the same input and experimental data with which Figure 4.17 was plotted, but calculated in a different way. To generate the SMA curve, the experimental input data is the daily flow of methane produced by the acetic acid (or acetate), instead of the cumulated volume of methane. If employing an AMPTS, the calculation is straightforward, because the instrument provides information of both the cumulated volume and daily flow of each reactor. Other instruments may require a bit more manual calculations, but the eventual additional effort is irrelevant. The general idea of the SMA test is similar to that of the BMP test: Subtracting the daily methane production of the blank reactor from that of the reference reactor introduces the variable *time*, i.e., one has a clearer idea of *how quick* the conversion of acetate to methane is. Since the VSs of the inoculum are assumed as a representative measure of the amount of living bacteria, dividing the daily methane flow by the VS of inoculum gives a figure, which shows *how active* the archaea population is. When plotting the result

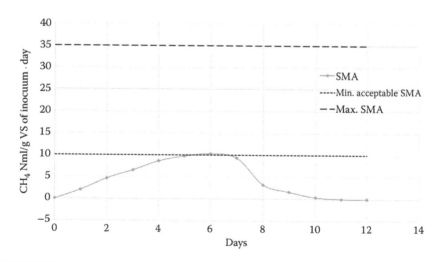

FIGURE 4.18

SMA test performed with wine vinegar as reference substrate and the same inoculum employed for the hydrolytic capacity tests shown in Figures 4.14, 4.15, and 4.17.

of said calculations, the curve has usually the shape of a bell. By definition, the maximum value (the peak of the bell) is the SMA.

There is a large disparity of criteria in the literature about how to calculate the peak SMA value. Some researchers check the hourly methane production rate, and extrapolate to 1 day the maximum hourly value (i.e., multiply by 24 the measured value, resulting in very high numerical coefficients). A few other researchers employ a mixture of acetic, propionic, and butyric acids (or their sodium salts) as reference substrate, which is scientifically more correct than using only acetic acid, but makes the test more difficult to perform in an industrial plant's laboratory. For the practical purpose of managing a biogas plant, we will always perform the simplified test (with acetic acid, sodium acetate, or just wine vinegar), adopting always the net daily methane flow as input data for the calculation. Under said rules, the acceptable values of the SMA will be in the range of $10–30 \, Nml/g \cdot VS_{inoc} \cdot day$.

Returning to the example of the inoculum with inhibition problems, when plotting the SMA curve, it is evident that the inoculum in question has an SMA value, which is at the lowest acceptable limit. This explains the anomalous shape of the curves in Figure 4.13. We can hence conclude that, in the case under study, there is no problem with the hydrolytic capacity; on the contrary, it is higher than the methanogenic capacity; hence sugars and starch are converted into VFA at a higher rate than the archaea are capable of metabolizing into methane, resulting in VFA accumulation. The acidification of the inoculum leads to partial inhibition of the archaea, which explains why the methane production curves of sugar and starch present "humps." Protein and cellulose, being complex substrates, have slower hydrolysis rates, so VFA are produced more or less at the same speed at which the archaea are able to convert them into methane, resulting in regular methane production curves (although a bit "flat" for a trained eye).

In the case of the biogas plant where the situation described was found, the owner had never been able to explain why the plant could run acceptably (although not without some troubles) when fed with corn and triticale silage, but when fed with sugar beet or molasses the production dropped instead of growing. The owner had spent a lot of money in buying a titrator and trying to manage the process on the basis of the FOS/TAC (VFA/TA) parameter, which showed no anomalies because of the inoculum's high alkalinity (dry cow manure), which masked the acidification. Sending samples to an external laboratory to analyze the VFA profile had proved of no help to solve the enigma, until the simple tests with reference substrates described earlier allowed to have a clear vision of where the biological bottleneck was.

4.4.2 Practical Example of the SMA Test for the Selection of the Inoculum Necessary for Starting a New Biogas Plant

Another application of the SMA test is the selection of the most suitable inoculum for starting a new biogas plant (or for restarting a plant after a biological collapse, or for "reinforcing" the bacterial ecosystem in a trouble-some plant as the one described in the former example). The following example, taken from a real case, shows how the inoculum from different biogas plants, although all of them running apparently without any trouble, may show very different SMA. In this case, it was necessary to start a newly built agricultural biogas plant. The options to inoculate it were:

1. Taking sludge from a nearby biogas plant that treated industrial effluents (inoculum A)

2. Taking sludge from an agricultural biogas plant fed with cow manure and corn silage (inoculum B)

3. Taking sludge from a biogas plant fed mainly with mixed silage, cow and pig manure from time to time (inoculum C)

The corresponding SMA curves are shown in Figure 4.19.

All three inoculums reach peaks higher than the minimum acceptable SMA ($10 \, Nml/g \cdot VS_{inoc} \cdot day$); nevertheless, inoculum B appears as the best one. The negative portion of inoculum B's curve is the result of a "quick test," i.e., the inoculum had not been degassed enough. Inoculum C is good, while inoculum A probably has some deficit of trace elements, because the SMA curve shows some ripple and reaches the peak a bit "late," on the 5th day.

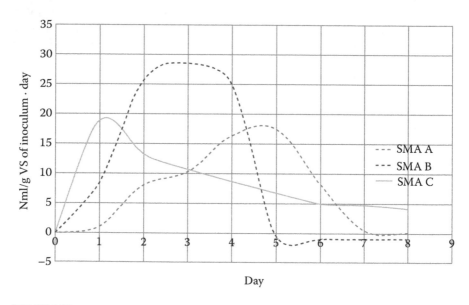

FIGURE 4.19
Different SMA curves generated by inoculums sampled from biogas plants running on different substrates.

4.4.3 How to Determine the Dosage of Trace Elements if SMA < 10

The same basic SMA test already described will be repeated, but this time with three additional reactors, each one added with increasing dosages of *medium*. Medium, a.k.a. *mother solution* or *micronutrients solution*, is an aqueous solution containing trace elements in a concentration 1000 times higher than the minimum, so that adding 1 ml of medium per liter of inoculum should provide the acetoclastic archaea with the minerals they need. Adding different dosages to the reactors will allow comparing the SMA of each one, and finding out the optimum, which may be higher than the minimum threshold. The recipe of the medium is provided in Table 4.3. The steps for performing the test are the following:

1. Sample the inoculum from the digester and sieve it through a 5 mm mesh. Measure its VS or its COD.
2. Fill four reactors with the same volume of inoculum, *V*, and label them "control," "1x," "2x," and "3x." Add 1 ml of medium per liter of inoculum to "1x," 2 ml/l to "2x," and 3 ml/l to "3x." A micropipette or a burette will turn very useful for correctly dosing the medium to each reactor.
3. Close the reactors, flush the headspace, and preincubate them at least 7 days at the same temperature of your plant. The scope of this operation is double: degassing the inoculum for higher test accuracy and allowing the micronutrients to be absorbed by the microbes, reactivating thus their enzymatic activity. It is advisable to check the biogas production until the cumulated volume curve becomes more or less flat or until the daily specific methane production is <10 Nml/g VS of inoculum·day.
4. Once the four reactors are degassed, add to each reactor 33 g of vinegar per liter of inoculum. It is not necessary to open the reactors—the

TABLE 4.3

Recipe of the Medium and Application to the Plant

Compound	Medium Concentration [g/l of Solution]	*i*	Digester's Volume [m³]	Quantity to Digester [kg] Medium × *i* × Volume
Cl_3Fe	75.5			
$NiSO_4$	5.0			
$ClZn$	0.1			
$CoSO_4$	0.9			
$MnSO_4$	0.3			
$NaSeO_4$	0.3			
$Mo_2(NH_4)_2$	0.4			

advantage of using vinegar instead of sodium acetate is that the former can be added with a syringe through the tube with valve used for flushing the headspace volume at the beginning of the test, without opening the reactor.

5. Start the experiment, wait at least 4 days, and check the daily flows to find the maximum daily biogas production of each reactor—in general, it will be reached within 48–72 h. We call this value Peak [Nml/day].

6. Calculate the SMA of each reactor with the following formula:

$$SMA_i = \frac{Peak_i}{V \cdot VS_{inoculum}}$$

The micronutrients dosage for the plant will be calculated from the results of the test described earlier. The ideal micronutrients treatment is the minimum dosage that produces the maximum effect. For instance, suppose that the calculated SMAs are the following:

$SMA_{control} = 6$ Nml/g VS·day

$SMA_1 = 11$ Nml/g VS·day

$SMA_2 = 20$ Nml/g·VS·day

$SMA_3 = 23$ Nml/g·VS·day

The $SMA_{control}$ value of this example is typical of a plant suffering biological collapse. The dosage $1x$ could suffice to bring it to a minimum activity, but the resulting SMA is just slightly higher than the acceptable minimum. The dosage $2x$ brings the plant to a "good" SMA. Dosage $3x$ produces very little improvement compared to dosage $2x$, so its application would be wasting money in overdosing. We will hence add to our plant's feedstock a cocktail of minerals to reestablish the methanogenic activity, having the same composition of the medium. The quantity of each mineral compound to be added to the digester (in kg) can be calculated as the concentration employed in the medium (in g/l), multiplied by i (where i = 1, 2, or 3 depending on the optimum SMA test found), multiplied by the volume of the digester in m³. Table 4.3 is self-explanatory.

Figure 4.20 shows the result of adding the calculated dosage to a 1-MW (electric) biogas plant that was inhibited from the lack of trace elements. The increase in output power was relatively steady after 5 days of the treatment.

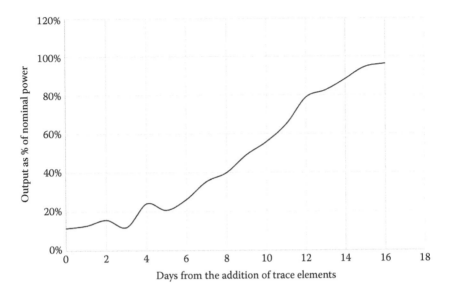

FIGURE 4.20
Effect of adding the correct dosage of trace elements to a biogas plant that presented impaired methanogenic activity.

4.5 Conclusions

4.5.1 Usefulness of the Test with Vinegar, Acetic Acid, or Acetate as Reference Substrate

The AD test using vinegar, acetic acid, sodium acetate, or a mixture of VFA provides two pieces of information that the biogas plant manager must evaluate *simultaneously*. The first is the *capacity* to convert all the substrate in methane (hence, the ultimate methane production compared to the theoretical BMP of VFAs). The second is the *maximum speed* at which the said conversion is achieved, referred to as the mass of living bacteria represented by the VS (i.e., the SMA). The SMA is useful both for periodically checking the health status of the digester's bacterial ecosystem and for selecting the most suitable inoculum to start a biogas plant.

4.5.2 Reference Substrates

The practical examples presented in Sections 4.3 and 4.4.1 demonstrate that some simple tests performed with cheap and widely available reference substrates provided more useful information than that obtained through expensive analysis performed by external "specialized" laboratories.

In conclusion, it is more important to know *what* and *how to* measure the real microbial activity and digestion efficiency, rather than deciding our plant management strategies on generic guides, tables from "the literature," and expensive tests performed by third parties.

Bibliography

Bolzonella, D., Monitoring and lab analysis—Assays, Notes on the course on biogas technology, Jyvaskyla Summer School, Jyväskylä, Sweden, 2013.

Burke, D., *Dairy Waste Anaerobic Digestion Handbook*, Environmental Energy Company, Olympia, WA, 2001.

Hutòan, M., Mrafkovà, L., Drtil, M., and Derco, J., Methanogenic and nonmethanogenic activity of granulated sludge in anaerobic baffled reactor, *Chemical Papers* 53(6), 374–378, 1999.

Lozano, C.J., Mendoza, M.V., de Arango, M.C., and Monroy, E.F., Microbiological characterization and specific methanogenic activity of anaerobe sludge used in urban solid waste treatment, *Waste Management* 29, 704–711, 2009.

Miranda, L., Henriques, J., and Monteggia, L., A full scale UASB reactor for treatment of pig and cattle slaughterhouse wastewater with a high oil and grease content, *Brazilian Journal of Chemical Engineering*, 22(4), 601–610, 2005.

Rosato, M., Quick restart of a biogas plant and micronutrients cost minimization, Application note of Bioprocess Control AB, http://www.bioprocesscontrol.com/company/media-gallery/application-note-quick-restart-of-a-biogas-plant-and-micronutrients-cost-minimization/, 2016.

Schneiders, D., da Silva, J.D., Till, A., Lapa, K.R., and Pinheiro, A., Atividade metanogênica específica (AME) de lodos industriais provenientes do tratamento biológico aeróbio e anaeróbio, *Revista Ambiente & Água—An Interdisciplinary Journal of Applied Science* 8(2), 135, 2013.

Van Lier, J., Sanz, M., and Lettinga, G., Effect of temperature in the anaerobic thermophillic conversion of granular and dispersed sludge, *Water Research*, 30, 199–207, 1996.

Zhao, Y., Ma, S., Sun, Y., Huang, Y., and Deng, Y., Isolation, identification and enzyme characterization of a thermophilic cellulolytic anaerobic bacterium, *Wei Sheng Wu Xue Bao*, 52(9), 1160–1166, 2012 (in Chinese).

5

Some Simple Tricks to Improve the Laboratory's Operativity

5.1 Foreword

Most of the techniques illustrated in this chapter were implemented and fine-tuned with an AMPTS II instrument, and the photos show the same, just because the author's laboratory has one. Nothing prevents that, and with the necessary adaptations, they can be implemented with other instruments too.

5.2 Measuring the Percentage of Methane in the Biogas with the Double Reactor Set and with the Syringe Method

In the academic world, the most widely diffused technique to analyze the chemical composition of biogas is gas chromatography. The said method is very accurate, but its cost and complexity place it beyond the possibilities of the average biogas plant manager. Almost all industrial biogas plants are equipped with a fixed online analyzer of the biogas composition, usually based on the infrared spectrometry (IRS) method to determine the proportions of carbon dioxide and methane. Portable biogas analyzers with IRS and/or solid-state sensors are also available in the market. Such instruments are not suitable for laboratory measures, since they require "big" volumes of biogas at each reading. For instance, the Geotech Biogas 5000 portable biogas analyzer requires a biogas flow of 500 ml/min, far beyond the capacity of any laboratory reactor. Sometimes it is important to check in the laboratory not only the net amount of methane produced by the samples, but also the quality of the biogas, i.e., its methane percentage. A simple way to obtain two individual curves, one showing the production of methane and the other the production of carbon dioxide, consists in measuring the difference between the daily production of biogas and the daily production of methane.

To perform this test, two identical sets of blank and sample reactors must be prepared. The first one must be connected to the instrument as usual, i.e., through the caustic soda filters, measuring the net methane production. The second set must be connected directly to the instrument, measuring then the total biogas production.

The elaboration of the data with the spreadsheet will be the usual—subtract the average of the blank reactors from the average of the correspondent sample reactors, and divide the said difference by the VS of the substrate. We will then obtain two columns in the spreadsheet, one containing the average daily specific production of biogas, and the other containing the daily specific methane production. It is easy to add a third column to the spreadsheet, containing in each cell the difference between the specific production of biogas and the specific production of carbon dioxide. Now it is easy to plot both curves, net production of methane and net production of carbon dioxide as a function of time. Alternatively, one can plot the percentage of methane as a function of time. Figure 5.1 shows an example of the results obtained with the test described here.

Observe that the CH_4 percentage measured in batch tests like the one shown in Figure 5.1 tends to be overestimated in comparison with real-life experience with the same substrates, especially at the end of the test. The reason is that not all the biogas produced by the inoculum is released and passes through the instrument. Indeed, some of the CO_2 is captured by the inoculum itself, forming carbonates and bicarbonates. In some extreme cases, for instance, in the SMA test using sodium acetate, the biogas composition resulting from the batch digestion may reach 90% of CH_4, because

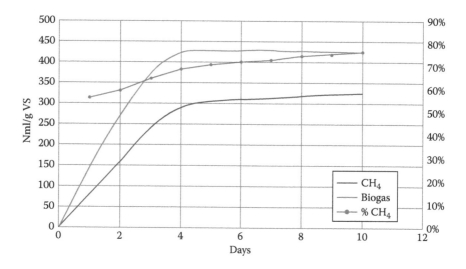

FIGURE 5.1

Comparison between the normalized curves of biogas and methane production, and the variation of methane percentage in the gas as a function of time.

the CO_2 is immediately captured by the NaOH that forms when the free Na, released by the archaea while digesting the $C_2H_3NaO_2$, reacts with the surrounding water and then with the CO_2. This is a feature of the batch tests, not a defect of the instrument or an error in the test conduction. The only way to minimize the said effect would be to saturate the sludge with CO_2 before starting, but such operation adds complexity to the test protocol and may alter the inoculum's buffer capacity. Figure 5.2 shows an example of the double reactor test with sodium acetate as substrate.

The comparison of the three curves in the same graph provides useful information about the kinetics of the substrate's degradation, since the proportions of methane and carbon dioxide vary with time according to the fermentation rate, and this is an important factor to consider in some plants, especially those with two or more stages. For sure, performing two sets of the same test in parallel, with and without carbon dioxide filters, is the most complete method to understand the dynamics of the AD process, since it provides *continuous* information about the hydrolysis and methanogenesis rates. On the other hand, such method requires employing double number of reactors for the same substrate. Users who do not possess an AMPTS II or similar, i.e., an instrument having many measure channels with real-time data logging and normalization, will have to perform more manual reading, normalization, and calculation so as to generate the curves described. In short, measuring in parallel total biogas and net methane requires some extra work and occupies reactors that we could otherwise employ for testing more substrates or for other biological activity tests.

The syringe method, described in the next paragraph, is very simple and acceptably accurate for discrete measures (e.g., once a day) of the biogas

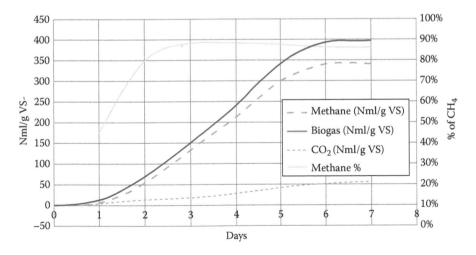

FIGURE 5.2
Specific production of raw biogas and net methane from sodium acetate, as measured with the double batch test. In this particular case, the percentage of methane in biogas can reach 90%.

composition, assuming the latter as a bicomponent mixture of methane and carbon dioxide. It can be employed for both batch and continuous tests, since it provides just points, for instance, one measure per day. The method is based on the great capacity of aqueous caustic soda solution to absorb carbon dioxide, and on *Avogadro Law*:

> The volume occupied by a mixture of two gases is the sum of the volumes occupied by the single gases at the same conditions of pressure and temperature.

The validity of this method is justified because the other components of biogas (ammonia and hydrogen sulfide) exist in concentrations of the order of parts per million, negligible if compared to the concentrations of methane and carbon dioxide, which are always around 50%.

5.2.1 Necessary Materials for the Syringe Test

- A special syringe for gas sampling, at least 15 ml capacity, with valve (Figure 5.3)
- 1 g of caustic soda diluted in 10 ml of distilled water
- A gas sampling valve with rubber septum (Figure 5.4)

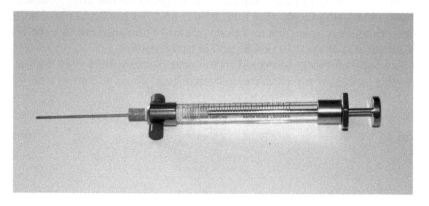

FIGURE 5.3
A special syringe for gas sampling, equipped with valve. (Photo by the author, 2-ml model manufactured by VICI, http://www.vici.com/syr/a2.php.)

FIGURE 5.4
Gas sampling valve with rubber septum. Courtesy of Bioprocess Control AB.

5.2.2 Performing the Test with the Syringe

1. Insert the needle in the septum, open the valve, and aspire 10 ml (or 1 ml, depending on the available syringe) of biogas. Wait a few seconds until the pressure reaches the equilibrium and close the valve. See Figure 5.5.

2. Submerge the needle in a small recipient containing caustic soda solution, open the valve, and aspire at least 10 ml (or 1 ml in this example) of solution. Close the valve and shake the syringe for a few seconds. The caustic soda will absorb the CO_2. See Figure 5.6.

3. Immerge again the needle in the caustic soda solution, open the valve, and push the plunger to its initial position (10 or 1 ml in this example). Do not close the valve, and allow some seconds for the pressure to stabilize.

4. While keeping the needle immersed in the solution, hold the syringe in vertical position and read the remaining volume of solution; see Figure 5.7.

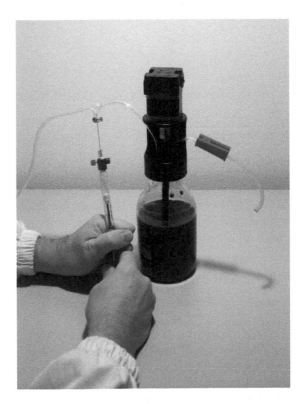

FIGURE 5.5
Aspiring the biogas sample. The green button shifted to the syringe body indicates the position "open." Press the red button before retiring the syringe from the sampling septum.

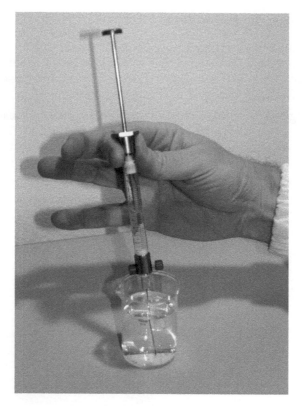

FIGURE 5.6
Syringe full with biogas and caustic soda solution.

The said volume of aqueous solution corresponds to the initial volume of carbon dioxide present in the biogas sampled, while the remaining volume of the syringe is now full of methane. In the example shown in Figure 5.7, the volume of solution is 0.30 ml; consequently, the biogas sampled from the reactor is composed of 30% carbon dioxide and 70% methane.

5.2.3 Error Analysis of the Syringe Method

As in all volume measure methods, the error sources are:

1. The error class of the syringe. The minimum reading (absolute error) of a well-trained eye is half the space between two divisions of the volume scale engraved on the syringe. In our example, the scale spacing is 0.5 ml, hence the absolute error of the reading is ±0.25 ml, corresponding to ±2.5% absolute error of the CO_2 percentage.
2. Human errors. The superficial tension of water surfaces forms a meniscus. When reading the volume of solution remaining in the

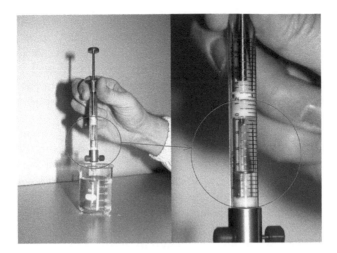

FIGURE 5.7
The syringe with the plunger pushed back to its initial position.

syringe, the latter must be kept perfectly vertical, and the reference line is the lowest point of the meniscus.

Any of the following strategies allow to improve the accuracy of the syringe method:

1. Sample a bigger biogas volume (e.g., employing a 50- or 100-ml syringe). This is feasible only if the reactor's head volume is big enough. A 500-ml reactor, as the one shown in Figure 5.5, does not allow aspiring more than 10 ml, otherwise the internal pressure would drop to the point of saturating the biogas with water vapor. The vapor would then condensate in the syringe, leading to the over-estimation of the carbon dioxide concentration.
2. If sampling the gas with a 10-ml syringe, discharge the remnant solution into a 5- or a 10-ml burette, class A, and use this to measure its volume. Such a burette has 0.05 ml graduations, with an accuracy of $\pm$ 0.025 ml, hence the absolute error in the CO_2 measure when sampling 10 ml of biogas becomes $\pm$ 0.25%.

5.3 Solving the Problem of the Thermostatic Bath's Evaporation

One of the most interesting features of the instruments for biological tests including a data logger, like the AMPTS II and similar, is the possibility to access the data through the Internet. This allows the biogas plant manager to launch the tests and let the instrument do all the work, so it is not necessary to stay all day in the laboratory. The drawback is that, if one does not check frequently the level of the water bath, it may evaporate until the reactors remain dry. In such case, the automatic protection will switch off the resistance and the reactors will cool, giving wrong biogas production curves.

The evaporation rate is higher when the laboratory's temperature is low (<20°C, as for instance, during weekends in winter); when the air in the laboratory is too dry (RH < 60%) or when conducting experiments under thermophilic conditions. If the volume of the water in the incubator drops to such a level that the reactor's temperature does not remain uniform, the gas production curves will show conspicuous variations of the fermentation rate that the instrument will detect. When refilling the water bath to compensate the evaporation, the incubator's temperature will vary abruptly. Such sudden temperature change reflects in the methane production curves as "steps" on the general sigmoid, as shown in Figure 5.8.

The solution to the problem just described consists in lowering the evaporative rate. The following paragraphs explain some of the possible strategies to adopt.

5.3.1 Seal the Gap between the Reactors and the Plexyglass Cover by Means of *O-rings* or Rubber Bands

This operation reduces the evaporative rate, but does not stop it, because it is impossible to obtain a perfect seal.

5.3.2 Add Some Very Soluble Salt to the Water

This trick is based on *Raoul Law*:

$$P_s = X_a \cdot P_0$$

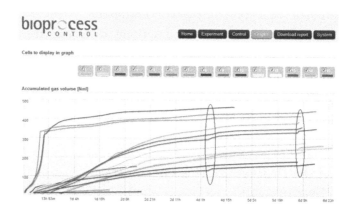

FIGURE 5.8

The steps on the general sigmoid curve are the consequence of temperature variations of the thermostatic bath, caused by the evaporation and subsequent addition of water.

where P_s=vapor pressure of the solution at the reference temperature; X_a=molar concentration of the water in the solution, in %; P_0=vapor pressure of pure water at the reference temperature

In other words, the vapor pressure of any aqueous solution decreases when the concentration of the solute grows, because the concentration of the solvent (water) decreases.

For instance, when carrying out an experiment at 40°C with the thermostatic bath full of distilled water, its vapor pressure, P_0, will be 7.37 kPa. Suppose that common salt, ClNa, is added to the water. The maximum solubility of ClNa at 40°C is 366.9 g/l. Considering that 1 l of water contains 55.55 mol and that 366.9 g of ClNa are equivalent to 6.32 mol, the water concentration in this saturated brine will be:

$$X_a = \frac{55.55}{55.55 + 6.32} = 89.7\%$$

Hence, the evaporation rate of brine will be 89.7% that of distilled water. In other words, the time to evaporate a given quantity of water from the brine will be only 10.3% longer compared to the same quantity of distilled water. By employing other salts, for instance, a saturated solution of LiBr (lithium bromide), the increase in time between two recharges of the water bath could reach 20%. The drawbacks of this solution are: triggering corrosion effects in the thermostatic bath when employing NaCl, high cost if employing LiBr or other similar salt, precipitation of salt in any case. The reduction of the evaporation rate provided by this method is not as high as one should desire, hence the use of brine is not an optimum solution for the problem.

5.3.3 Replacing the Water in the Thermostatic Bath with Any Fluid Having Low Vapor Pressure

The best commercial candidates for this scope are silicone oil, glycerol (glycerin), ethylene glycol (EG), and propylene glycol (PPG). Silicone oil is expensive and potentially toxic, with the additional drawback of its lubricating quality that would make the reactors difficult to manipulate once an experiment is finished and another is started. Glycerol has an oily texture too, but its toxicity is low (it is used as sweetener in the food industry and as skin moisturizer in the cosmetic industry). Glycerol being biodegradable and considering the favorable temperature at which the incubator usually works, there is a risk that aerobic bacteria could alter its composition in time, so it is not convenient for our scope. EG and PPG are both cheap and widely available. EG is available at any petrol station, or big supermarket, or DIY shop, since it is one of the most diffused antifreezing products for car radiators. The toxicities of both EG and PPG are low—slightly better PPG from this point of view—but none of them are easily degradable by aerobic bacteria.

The vapor pressures of PPG and EG at 40°C are 0.045 and 0.039 kPa, respectively. Compared to 7.37 kPa of distilled water at the same temperature, it means that both EG and PPG will evaporate about 200 times more slowly than water.

The relationship between temperature and vapor pressure of the mixture water/EG is shown in Figure 5.9.

The thermal capacity of EG is lower than that of water (2.404 kJ/kg·K for EG, compared to 4.181 kJ/kg·K of water). The thermal conductivity of EG is about half that of water, but enough to maintain the uniformity of temperature in the incubator, where the convective thermal exchange is stronger than the conductive one. The low evaporation rate of EG constitutes another advantage of this liquid on water: the energy consumption of the incubator will be smaller because the latent heat dissipated with the evaporation will be negligible. The author used to consume up to 5 l of distilled water each week in a thermostatic bath working at 38°C. Since he replaced water with EG, the evaporation became almost imperceptible. This represents an energy saving equal to:

$$E_{evap} = 5\,kg \cdot 2,559.1\,kJ/kg\,week = 12,795.5\,kJ/week = 14.21\,kWh/month.$$

Other advantages of EG, that make it preferable to water, are the absence of calcareous scale, unavoidable when employing tap water, and of rust spots,

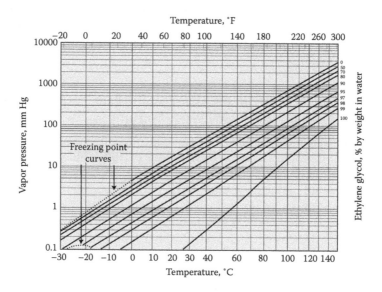

FIGURE 5.9
Vapor pressure of water/EG solutions as a function of temperature. Taken from the technical sheet of EG published by the producer MEGlobal.

caused by the acidity of demineralized water, since it contains dissolved carbon dioxide.

An advice if you intend to purchase EG at a petrol station or in the car supplies sector of a big supermarket: read carefully the chemical composition in the label, because there are two different types of antifreezing liquids for radiators. One of them is a mixture of water/alcohol (often colored in red, but this may change from country to country). The said mixture is a good antifreeze, but it is not suitable for our purpose, since it evaporates quicker than water if left to open air. The second is EG, usually colored in blue or green, often containing water in different percentages (cheaper products contain more water and must be avoided; search for the product with the highest purity possible). During the first days running a test with the thermostatic bath full of glycerol, you may observe that the level drops. Add more product until all water will have evaporated and the liquid level in the thermostatic bath will remain stable, just below the plastic lid.

5.4 Improved Connection of the DC Stirrer Motors

5.4.1 Description and Theoretical Analysis of the Problem

Some laboratory reactors are equipped with stirrer motors, while others are not stirred, or rely on just magnetic bar stirring. Stirring is an important parameter when working with dense sludge (more than 5% VS), like the one in agricultural biogas plants. Stirring motors are not all the same—some are of the AC squirrel cage type, others are fed with low voltage DC, the most modern ones being of the brushless type. If performing many tests simultaneously (the complete rational management of an industrial or agricultural biogas plant requires at least six reactors), one may often observe that the rotation speed of the stirrers is not uniform. This becomes very evident when running low voltage DC motors from a single power source, as was the case in the old version of AMPTS II, having 15 DC stirrers fed at 5 V from the same source. The problem becomes especially annoying when the sludge is very viscous, since the last motors of the array do not receive enough power and turn very slowly; in some case, they do not turn at all. The reason is that the stirrers are all connected parallel, as depicted in the scheme of Figure 5.10., through rapid connectors called *fastons*, inserted in the bushings of each motor.

Note that with such connection system, the DC motors are indeed in parallel with each other, but the contact resistances of the connectors are in series with the main flow of the current, as can be deduced from the equivalent electric diagram in Figure 5.11.

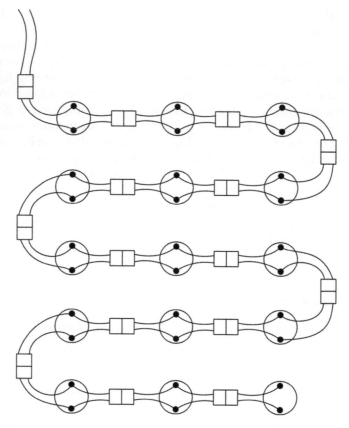

FIGURE 5.10
Pictorial scheme of the connection in parallel of several stirrers, in particular, the old version of AMPTS II.

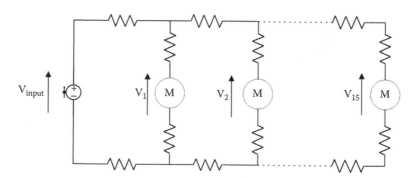

FIGURE 5.11
Equivalent electric circuit of the parallel connection of several direct current (DC) motors, considering the contact resistance of the *faston* connectors. Each resistance represents each single contact between two wiring elements.

With such a connection, each motor will receive a fraction of the voltage applied to the one immediately precedent. The power at the shaft of small DC motors is 50% of the absorbed electric power, and the latter is the product between voltage and current. Since the voltage drops in each connector between motors, and in the contact at the motor's bushings, the available power decreases along the connection cable, with the last motor receiving sometimes half of its nominal voltage or less. Another feature of DC motors is that their torque is proportional to the square of the current, and this latter is inversely proportional to the rotation speed. Hence, if the sludge in one of the reactors is very viscous, the motor will turn slowly and absorb more current. More current means more voltage drop across the contact resistances, so all motors downstream will receive less voltage, hence less power, hence will turn more slowly, the last of the line arriving sometimes to a complete stop. In spite of being at a stop, the current continues to circulate through the motor's windings, eventually overheating them. In the case of the AMPTS II, the control software is able to limit or switch off the current, to avoid any damage, but this means that all reactors will remain without stirring until the operator notices that there is a fault.

Table 5.1 shows the voltage at each motor's connectors when the AMPTS II stirring is set at 50% (5.52 V at the connectors of the first motor of the line)

TABLE 5.1

Voltage at Each DC Motor Connector Along a Line of 15 Motors Connected in Parallel by Means of *fastons*, with the Power Supply at One of the Line's Ends, as Shown in Figure 5.11

Voltage No.	Parallel Connection, Power Supply at One of the Line's Ends (V)
V_1	5.52
V_2	5.34
V_3	5.16
V_4	5.13
V_5	4.93
V_6	4.89
V_7	4.70
V_8	4.66
V_9	4.63
V_{10}	4.61
V_{11}	4.55
V_{12}	4.50
V_{13}	4.47
V_{14}	4.32
V_{15}	4.31

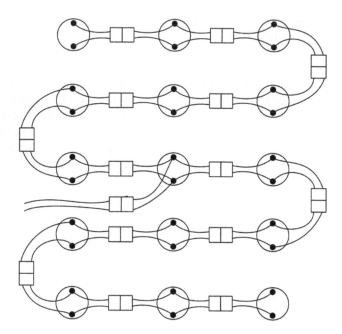

FIGURE 5.12
Connection of DC motors in parallel through the center of the power line.

and the reactors are loaded with a thin sludge, containing less than 5% VS. Observe that even under such favorable conditions the last stirrer of the line receives only 4.31 V, i.e., 78% of its nominal voltage. The rotation speed at the shaft of stirrer no. 15 is much smaller than that of stirrer no. 1, and the difference can be perceived by the naked eye.

The solution to the said problem consists in connecting the motors according to the scheme shown in Figure 5.12.

Under the same test conditions adopted for the elaboration of Table 5.1, if the power supply is connected at the center of the line, the voltages across each single motor increases notably, as shown in Table 5.2.

The difference of the voltage distribution between both connection systems can be appreciated in more detail in Figure 5.13.

5.4.2 Step-by-Step Procedure to Connect the Power Supply at the Center of the Line

First, note that the connectors of the cable elements are of two types—one end of the cable has a pair of female *fastons*, while the other end has a pair of male–female *fastons* (Figure 5.14). The motors instead have always male *fastons* (Figure 5.15).

TABLE 5.2

Voltages Across Each Motor Along a Line of 15 Units
Connected in Parallel and Fed from the Center of the Line

Voltage No.	Power Supply Connected at the Center of the Line, at the Bushings of Motor No. 8 (V)
V_1	5.30
V_2	5.32
V_3	5.34
V_4	5.42
V_5	5.47
V_6	5.55
V_7	5.59
V_8	5.64
V_9	5.58
V_{10}	5.53
V_{11}	5.49
V_{12}	5.41
V_{13}	5.36
V_{14}	5.35
V_{15}	5.34

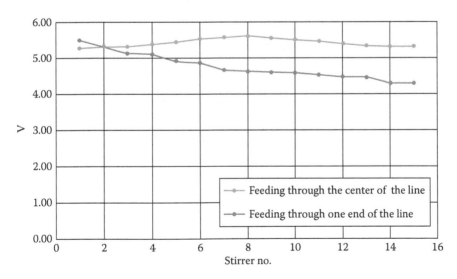

FIGURE 5.13
Comparison between the voltage distributions at each motor when connected in parallel, with
the power supply connected at one end of the line and in its center.

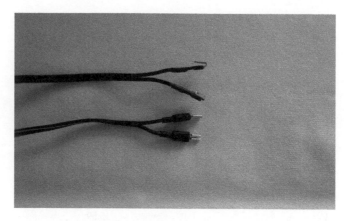

FIGURE 5.14
Detail of a connection cable for groups of DC motors in parallel.

FIGURE 5.15
Detail of the typical connection bushings (tabs) of 5 V DC motors usually employed for the stirring of laboratory reactors.

The following paragraphs describe step-by-step the connection procedure.

1. Insert the male–female connector into the tabs of the central reactor (no. 8 in AMPTS II and nos 3 or 4 in AMPTS Light), as depicted in Figure 5.16.
2. Insert another cable with male–female connectors in the free tabs of the connector in Figure 5.16, as shown in Figure 5.17.

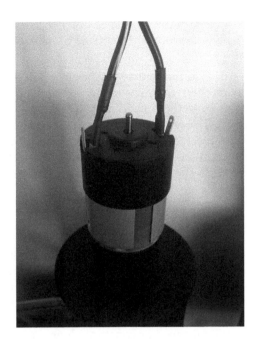

FIGURE 5.16
First step.

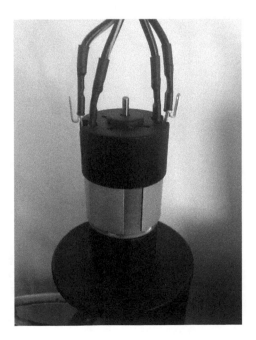

FIGURE 5.17
Second step.

3. Now, connect a cable with female fastons to the male fastons in Figure 5.17. The connection of the stirrer no. 8 should look now as shown in Figure 5.18.

4. Now that the stirrer of the central reactor (no. 8 in our example) has three cables with their corresponding connectors, connect one of them to the cable from the power supply (the long cable in the case of the AMPTS II). The remaining cables will be connected to those coming from the reactor at the right (no. 9 in our example) and to the reactor at the left (no. 7). From this point, connect the remaining reactors according to the usual male–female scheme. The array will look as shown in Figure 5.19.

5.4.3 Using Brushless Motors

Brushless motors are a particular type of DC motors, whose rotor consists of one or more permanent magnets, and the windings of the stator are fed with current pulses, generated in a precise sequence by an electronic driver. They can present almost constant torque in a given range of speeds, or torque that is inversely proportional to the rotational speed. Hence, their power can be variable with the rotational speed or constant along a given range of speeds,

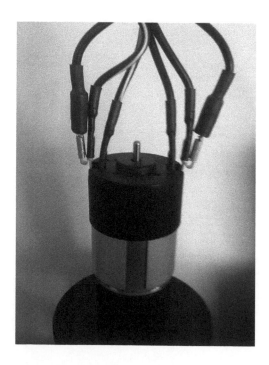

FIGURE 5.18
Third step.

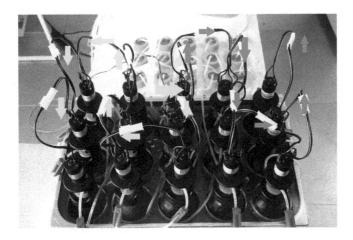

FIGURE 5.19
Overview of the connections of the 15-reactors array. The blue arrow shows the cable from the power supply, the green arrows show the branch from reactor 8 to reactor 1, while the orange arrows show the branch from reactor 8 to reactor 15.

FIGURE 5.20
Kit of stirring control up to 15 reactors with brushless technology. From left to right: 24 V DC power supply, electronic control unit, connection cables, aluminum nipple to connect the motor's shaft with the stirring shaft, and reactor cap with brushless motor. Photo by courtesy of Bioprocess Control AB.

depending on the driver's features. They are ideal to stir viscous sludge, since in general it will be possible to set a low rotational speed and keep it constant for the whole duration of the test. Eventually, the stirring power transmitted to the sludge can be modulated by means of *on/off* cycles. Figure 5.20 shows an example of such stirrers.

5.5 Checking the Calibration (Volumetric Methods)

We call *calibration* the operation by which a measure instrument is config-ured or adjusted so as to improve its accuracy. The operation requires com-paring the measures taken with the instrument under calibration with those taken with a *standard instrument*, or measuring magnitudes of known value, called *calibrators*.

The periodic calibration is necessary for some instruments that tend to lose accuracy with time (e.g., optical or mechanical measuring systems sub-ject to dirt and wear, electronic pressure sensors subject to thermal drift...). In the case of eudiometers and self-built liquid displacement instruments, it is advisable to perform a calibration over the whole measure range before starting to use them. The said extended calibration is useful in accounting for the compressibility of the gas, because the height of the water column is not always constant, and some of the water contained in the graduated cylinder may evaporate, giving a false reading. Commercial liquid displace-ment instruments working at constant water column (e.g., AMPTS, μ-Flow, MilligasCounter, and similar) are calibrated in their factory and in general do not require periodic recalibrations.

When calibrating instruments for measuring the volume of biogas (or of net methane) produced by AD, two possible methods are available: the gravi-metric method and the volumetric method.

5.5.1 Gravimetric Calibration Method

In this case, a given water volume injected in an airtight container dis-places equal gas volume (usually air). The said gas volume will be our standard magnitude, so we will then adjust the instrument so that it gives the corresponding reading. The sample volume of gas is measured by *weighing* the quantity of distilled water injected into the container. Hence, the standard instrument is an analytical scale, a kind of instrument usu-ally featuring high precision and accuracy. Remember that 1 ml of dis-tilled water corresponds to 1 g only when the distilled water is at 3.98°C. Consequently, to perform a correct calibration it is necessary to measure the temperature of the distilled water and to multiply the scale's reading by the corresponding specific volume, presented in Table 5.3. (it is accept-able to interpolate linearly for intermediate temperatures). Figure 5.21 shows the assembly scheme of a "field" gravimetric calibration system. A good quality moisture analyzer scale, or a scale with at least 1-mg accuracy could be sufficient as standard instruments. The recipient with any eventual initial quantity of water should be placed on the weighing pan and the scale should be then brought to zero. After connecting with tubes (as short as possible) the instrument under calibration and the water

TABLE 5.3

Density of Distilled Water at Different Temperatures

Temperature (°C)	Density (g/cm³)	Specific Volume (cm³/g)
0	0.9998395	1.00016
4	0.9999720	1.0000280
10	0.9997026	1.000297
15	0.9991026	1.000898
20	0.9982071	1.001796
22	0.9977735	1.00223
25	0.9970479	1.00296

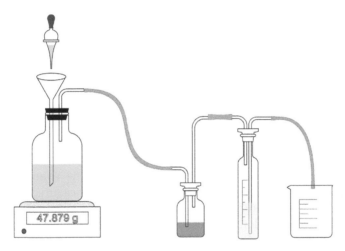

FIGURE 5.21
Scheme of a "field" gravimetric system to check the accuracy of a gas volume measuring device.

recipient, water can be injected in the latter by means of a syringe, or a burette, or a drop counter.

5.5.2 Volumetric Calibration Method

In this case, a known volume of distilled water, measured with a burette, displaces an identical volume of gas. The burette is then the standard instrument, less accurate than an analytical scale, but cheaper and straightforward to employ. Figure 5.22 shows the assembly scheme of a volumetric calibration system.

In general, a good calibration of an instrument includes preparing an error table. For instance, when calibrating the liquid displacement system shown in Figure 5.22 by means of a 10-ml burette having 0.1-ml divisions, the procedure

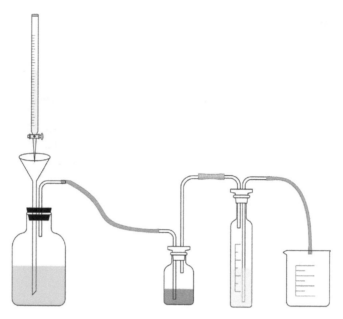

FIGURE 5.22
Assembly scheme of a volumetric calibration system.

consists in adding exactly 10 ml each time, and reading the volume effectively measured by the instrument under calibration. In the example shown in Figure 5.22, the displaced volume of water is usually measured with a graduated cylinder, having minimum reading (the space between graduations) equal to 1 ml. We must log the measured values in three columns: the cumulated volume injected with the burette in the first one, the volume measured by the instrument under calibration in the second, and the relative error of the reading in the third column. Thus, we obtain the distribution of the errors as a function of the reading. The maximum relative error found throughout the whole measure range will be the value that we must assume as instrumental error for the error propagation analysis. Subsequently, the normalization error must be added to the said maximum error of the measured volume. Eventually, the overestimation error of the reactor's head volume, if filled with inert gas at the beginning of the assay, must be considered too. The normalization relative error is equal to the sum of the relative errors of both ambient barometer and thermometer. The overestimation error caused by the moisture content in the measured gas and by the presence of an inert gas, is much more difficult to assess. We can estimate it as 5% of the total cumulated gas volume in the usual 30 days, if the blank and sample reactors have different head volumes. Since the said error belongs to the systematic type, if the blank and the sample reactors have the same head volume, the difference between both measured gas volumes will have no overestimation error.

5.6 Flushing the Head Volume of the Reactor with a Gas Lighter Recharge

Before launching a batch test, it is a usual procedure to purge the head volume of the reactor with an inert gas, to eliminate oxygen and allow the anaerobic digestion to start immediately. The most commonly employed flush gas is N_2, although some researchers defend using a mixture of 60% N_2 and 40% CO_2. Other inert gases, like He, can be employed too. Purging with pure CO_2 is not advisable when the reactor is connected to a filter with caustic soda, because the latter will quickly absorb the gas, creating a partial vacuum in the head space. Such partial vacuum could suck caustic soda back to the reactor, compromising or at least altering the result of the test. Another undesirable effect of the vacuum created under said conditions is the saturation of the headspace with water vapor, which may condense in the tubes and dilute the caustic soda solution.

In general, commercial biogas plants do not keep N_2 cylinders in their warehouse. Is it mandatory to purge the headspace with an inert gas? The answer is no. The reason why research laboratories employ N_2 or He is just because of safety reasons. In research laboratories, it is common that several researchers share a space, where it is possible that somebody is employing an open flame, like that of the Bunsen burners, so flushing with inert gases eliminates the risk of accidental fires. In a biogas plant, the situation is the opposite—there cannot be open flames; one is usually alone in the lab, and there is plenty of biogas that can be taken from the digester. Suitable containers for sampling enough biogas for flushing the reactors can be: a gas balloon, a "maxi" gas syringe (much comfortable to use but expensive), a capsized laboratory bottle with water seal. Aspiration hoods are not usually present in biogas plants, as is the case in research laboratories, but in general, there is no problem in opening a window or ensuring good ventilation with a small air extractor in a plant's laboratory. Furthermore, the head volume of laboratory reactors is very small, so even if purging them with biogas, or methane, or other combustible gas, it is physically impossible to create an explosive atmosphere.

The author usually purges the reactor's headspace with a gas lighter recharge. Lighter recharges usually contain propane, or a mixture of propane/butane, both hydrocarbons similar to methane, perfect for creating an anoxic condition. Lighter recharges are small cans with a safe valve, easy to handle, and available at any tobacconist's for a few dollars. We must remark that the headspace is initially full with air, which contains just 21% oxygen, the rest being nitrogen. Hence, if we flush the headspace with twice to five times its volume of propane, the proportion of oxygen remaining after the purge will be so low, that it will not harm or delay the digestion process. From the safety point of view, if a quantity of propane in the range of 200–500 ml diffuses in the air of the room, there is absolutely no risk of explosion.

The *lower explosion limit (LEL)* of propane is 2.1%, so even in the case of a very small room lacking any sort of ventilation, for instance, 9 m² with 2.4 m height, at least 450 l of propane would be necessary for forming an explosive mixture. Considering that lighter recharge cans contain around 42 g of butane, which means 16.5 L (density=2.5436 kg/Sm³), one should fully discharge into the room's atmosphere the content of 27 such cans before reaching dangerous levels.

Figure 5.23 shows a gas lighter recharge, available at the tobacconist for a few dollars, being employed to purge a laboratory reactor's head volume.

After flushing the reactor's headspace, but before starting the test, the laboratory operator must check his/her instrument's settings: incubation temperature, stirrer's speed, resetting the gas measure unit, etc. Some instruments like the AMPTS II and AMPTS Light, Gas Endeavour, and Bio Reactor Simulator, provide the possibility to activate an algorithm that accounts for the eventual partial vacuum caused by the absorption of carbon dioxide by the caustic soda solution. If employing biogas from the plant to flush the head volume of the laboratory reactors, it is necessary to define its CO_2 concentration together with the other instrument's settings. If employing any inert gas or butane from a lighter recharge, then it is necessary to set 0% in the corresponding instrument's software screen (Figure 5.24).

FIGURE 5.23
Flushing the air from a laboratory reactor with a gas lighter recharge.

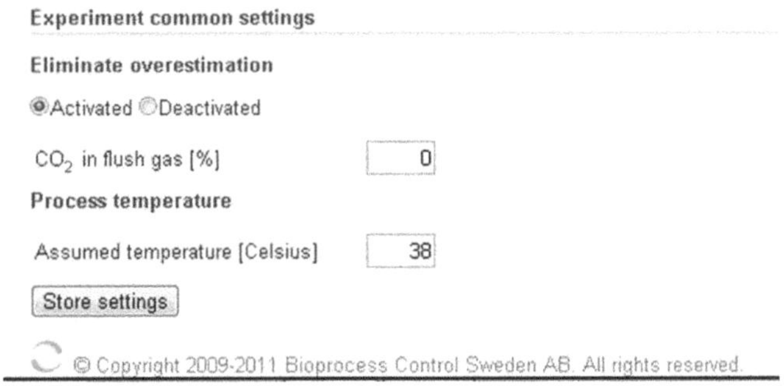

Experiment common settings

Eliminate overestimation

◉Activated ◯Deactivated

CO_2 in flush gas [%]   0

Process temperature

Assumed temperature [Celsius]   38

[Store settings]

FIGURE 5.24
Setting the initial experimental conditions in an instrument equipped with an algorithm to eliminate the overestimation of the measured gas volume. (Courtesy of Bioprocess Control AB.)

The simplest option when the reactor's headspace is small (100–200 ml) in comparison with the quantity of methane that the substrate will presumably produce (at least 1500–2000 Nml in a 500-ml reactor), consists in just not flushing the head volume at all. For instance, considering that in 100 ml head volume only 21 ml is oxygen (the rest is nitrogen), this oxygen will be converted in an equivalent volume of carbon dioxide by the facultative bacteria. The CO_2 will be immediately absorbed by the caustic soda solution; hence the test will start with a light depression and a modest increase in the gas' moisture, which will eventually condensate in the same reactor's headspace or in the filter with caustic soda. At the end of the test, all the head volume will be full with methane, hence the total volume of gas effectively counted by the flow meter will be:

$$V_{\text{measured}} = V_{\text{produced}} - 21\,\text{ml}$$

The error introduced by not flushing the air is hence 21 ml/1000 ml, i.e., 0.21%. Should the reader decide not to flush the headspace at the beginning of the test, and in the case he/she is employing an instrument with overestimation compensation algorithm, like the already mentioned AMPTS II and similar ones, the option "eliminate overestimation" shown in Figure 5.24 must be turned off.

5.7 Finding Gas Leaks

Gas leaks are a frequent cause of errors. According to the IWA draft of norm on BMP assay (see Chapter 6), it is advisable to check and identify eventual gas leaks in all reactors before starting any test. It is especially difficult to identify gas leaks when employing volumetric gas measure methods, because these work at near-ambient pressure, preventing the use of soap bubbles or similar solutions to find out the leak points. The following pictures show the procedure step-by-step in the case of a 500-ml standard batch reactor by Bioprocess Control. The present method, with any eventual adaptation, is suitable for both barometric and volumetric measure systems.

First step: Procure a syringe with a volume at least half of the reactor's headspace. Pull the piston up to its course end to fill the syringe with air, open the flush gas valve in the reactor (red clamp in the photo) and connect the syringe to it (Figure 5.25).

Second step: Choke the tube that connects the reactor with the measure system, either by hand or with the aid of another laboratory tube clamp. Push the syringe's piston to the end—or as strongly as you can—and hold it for at least 60 s (Figure 5.26).

Third step: Release the piston and wait some seconds. It should slowly return as the air in the headspace expands back to its original pressure (Figure 5.27). If it does not return, then there is some gas leak. Check or replace the reactor's cap, or the connection tubes or the clamp(s) and repeat the test. N.B. It is

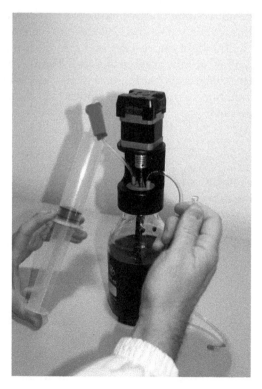

FIGURE 5.25
Setup of the test ready to start.

FIGURE 5.26
Pushing the syringe's piston to the end so as to pressurize the reactor's headspace. Ensure that the tube clamp in the flush gas port of the reactor (red plastic clamp in the photo) is open, and that the exit tube is correctly choked (eventually place an additional clamp on it and close tightly).

unlikely that the syringe's piston returns exactly to its initial position, because of the friction between the syringe's walls and the piston's rubber head. Please observe the difference of the piston positions in Figures 5.25 and 5.27.

Fourth step: Repeat the same procedure with the CO_2 traps (if you are employing a system equipped with them) and eventually with your gas measure device.

Hints:

1. When checking barometric measure sets, it is enough to pressurize the reactor, write down the initial pressure, and check it after some minutes. If the pressure decreases, then repressurize the reactor and apply soapy solution with a small paintbrush to all part joints. Small soap bubbles will show the points from where the gas leaks.

2. If the volume of the available syringe is much smaller than the reactor's headspace, then one piston stroke will not rise the pressure enough to overcome the friction when the piston is released, giving

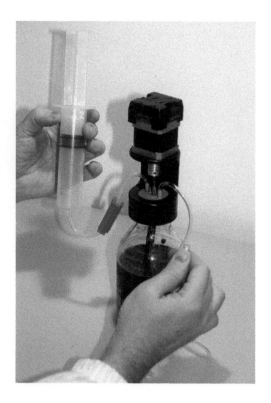

FIGURE 5.27
After releasing the syringe's piston, it will slowly return because the gas initially compressed in the reactor's headspace acts like a spring. In this example, the friction made the piston to return to about 70 ml instead of the initial 100 ml. This is absolutely normal—had there been a leak, the piston would not have returned when released.

the false impression that there is a gas leak in the reactor. In such case it will be necessary to pressurize the headspace by steps. Aspire air, then connect the syringe to the valve, open the valve, push the piston to the end, close the valve, retire the syringe, aspire air, and repeat the cycle for as many times as necessary so as to have enough pressure in the headspace.

5.8 Some Safety Rules for the Biological Laboratory in the Biogas Plant

Performing the simple biological and physicochemical tests described in this book does not create any particular risk beyond those usually encountered during ordinary housekeeping domestic work. The following safety advices are not prescriptions from any norm or regulation, but just common sense.

1. Keep domestic animals and children away from the laboratory.
2. Always wear disposable latex gloves and eye protection. This is mandatory when handling caustic soda, or acids, or sodium hypochlorite (bleach).
3. Caustic soda solutions should always be prepared with distilled or deionized water, adding the soda gradually to the water and stirring all the time. Avoid using tap water, since it will induce some precipitation (mostly carbonates). Adding too quickly caustic soda to water will most likely produce fumes from it. The same contain aerosolized sodium hydroxide, because the dissolution is exothermic and may give rise to microbubbles of water vapor. Avoid breathing such fumes, since they are extremely corrosive.
4. Always work in a well-ventilated room. If an aspirating hood or an air extractor are not available, at least open a window or a door.
5. When working with the muffle oven, the crucibles or other objects will be at 550°C, which is more than double the usual temperature of a kitchen oven. Always handle the hot objects with a special laboratory tong, whose length must be such that the operator's hands remain outside of the muffle's cavity (Figure 5.28).
6. Do not smoke or eat in the laboratory.
7. Do not fill beverage bottles with potentially harmful liquids like caustic soda solutions of acids: intoxication and burning risk by mistake.

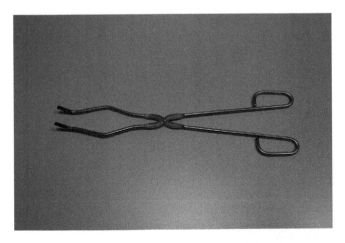

FIGURE 5.28
Special tongs for handling crucibles and other hot objects from the muffle oven.

8. Abundant fresh clean water, or a washing basin with an easy-to-open tap, must always be at hand for emergency washing in case of accidental contact with caustic soda or acids.

9. Avoid laying cable extensions on the floor: risk of stumbling upon them.

10. Always keep your work area clean and tidy.

11. A small blackboard and a piece of chalk often prove handy for writing down an instrument's reading or a memo, especially when one is wearing gloves or has dirty hands.

12. Turn off or mute your cellphone or radio while preparing a test. Distraction sources usually lead to human errors (e.g., forgetting to take the scale before weighing, exchanging two reactors, etc.).

13. Do not flush the digestate resulting from the tests in the water closet (WC). Although some digestate will not harm the sewage system, it is more sustainable to use it as fertilizer for the garden, or to dump it in the same plant's premix tank.

14. The exhausted NaOH solution of the CO_2 filters usually contains some active soda, so it can be employed to wash and disinfect the reactors and tools—except aluminum objects—and then it can be flushed in the WC (it is like the products sold in the supermarket for unblocking drain pipes). It can also be dumped in the premix tank, since it will just add some alkalinity to the sludge.

15. If using NaOCl (bleach) to wash and disinfect reactors and tools (not mandatory, but some people prefer to), just follow these simple rules: leave the objects overnight in a solution of one teaspoon (5 ml) of NaOCl per liter of water. After use, the solution can be disposed off in the WC.

16. A dishwashing machine, even an old one, turns very handy in the lab to wash glassware, spatulas, and similar tools. Bottles require washing by hand with a brush.

Bibliography

Bioproces Control, *Bioreactor Agitation Systems—Operation & Maintenance Manual*, Lund, Sweden, 2015.

Japan Soda Industry Association, *Safe Handling of Caustic Soda*, November 2006.

MEGlobal, *Ethylene Glycol Product Guide*, 2013.

Nistor M, *Calibration Protocol for the Gas Volume Measuring Device from AMPTS II*, Manual for internal use of Bioprocess Control Sweden AB, Version 1.1, August 2011.

6

Critical Review of the Scientific Literature from the Biogas Plant Manager's Perspective

6.1 Introduction

The goal of the present chapter is to help both biogas plant managers and academic researchers. The former will find a useful method to determine if information and data found in scientific journals can be applicable to their plants, while the latter will find some examples of conceptual errors or misconceptions that should be avoided or at least considered before publishing research results in peer-reviewed journals.

6.2 Three Methods to Find Out Absolute Truths: The Aristotelian Syllogism, the Cartesian Doubt Principle, and Avoiding Logic Fallacies

Biogas plant managers, especially when the plant's main commercial scope is energy production, are usually under pressure to increase the profit. This implies avoiding any modification to the routine that can potentially upset the process and lose profitability, but at the same time, there is a lot of pressure to find new ways of getting more methane per ton of feedstock and/or per cubic meter of plant capacity. It is in human nature to be afraid of the unknown, but making things always the same way brings no improvement. We have seen in the previous chapters how to apply Lord Kelvin's scientific method—"To measure is to know, and if you don't know something, you can't improve it"—to the biogas plant management, with the help of some simple, yet scientifically sound techniques. Since Internet offers tons of research papers to download with a few clicks, often free or at an affordable cost, the temptation to save labor and time by just implementing "discoveries" made by universities and other research bodies is high. Applying published research data to a biogas plant, without checking first in the laboratory the

suitability of such data, has a high risk of causing some damage to the anaerobic process because of the following reasons:

1. There is no warranty that the inoculum and test conditions employed
   for the published research are comparable to those of the biogas
   plant, hence trying to optimize a biogas plant on the basis of a few
   published tests can yield the opposite result and cause productivity
   and profit losses.

2. Error analysis seems to be a kind of taboo for the academic world
   researching on biogas, in spite of the consistent and well-proven set
   of ISO norms on error analysis and correct expression of the uncertainty of measured magnitudes. Quite often—practically always in
   the literature about anaerobic digestion—scientific papers do not
   include the analysis of *measure uncertainty* but just the *dispersion* of
   a given number of replicate tests. Sometimes papers do not even
   contain a thorough description of the instruments employed, and in
   some cases the resulting gas volumes are not even normalized. So,
   one cannot be sure if the "improvement" published by the researcher
   is really such, or just a combination of measure errors that led to a
   measured value higher than previously published ones.

3. Since the job of a researcher consists in creating a simplified model
   of a complex reality, sometimes the simplification is too much and
   the resulting model is valid only under very restrictive conditions.

4. Research institutions, even public ones, need money in order to carry
   on their activities. Quite often, sponsor companies fund research laboratories with the goal of developing or improving a given product.
   While such approach is legitimate, there is a strong probability that
   marketing claims present the research results in a biased or even dis-
   torted manner. The most common marketing hype is generalizing to
   all biogas plants the results obtained with just one particular case.

5. Quite often the "test results" presented as proof of the efficacy of a
   given product or method are based on the *post hoc* logical fallacy that
   will be explained in detail in the next paragraphs.

Adopting any new method or technique or feedstock in one's own biogas
plant requires a preliminary selection process based on the following:

1. *Reputation of the source*: It is clear to everybody that a paper published
   by a university or research institute is more likely to be scientifically
   correct than information published in a blog, even though the publication by a reputed source is not an absolute warranty of truth or
   correctness, or applicability to a particular case.

2. *Applying correctly the Aristotelian syllogism*: Syllogism (from ancient
   Greek συλλογισμός, syllogismòs, compound word of σύν, syn,

"together, joint, linked," and λογισμός, logismòs, "calculation, reasoning": hence "linked reasoning") is the base of logics and hence of the scientific method. Given two assumptions, any conclusion derived from these *can be true* only if both are true. The difference with digital or Boolean logics is that in real life there is no warranty that a conclusion derived from two true suppositions is *always* true. In logic terms, driving a conclusion from two true hypotheses is a necessary condition but not sufficient. On the other hand, a syllogism based on a false assumption may lead to an apparently true conclusion but just by pure coincidence.

3. *Applying the Cartesian doubt*: Also known as methodological skepticism, this method consists of four steps:

   a. Accepting only information you know to be true;

   b. Breaking down these truths into smaller units;

   c. Solving the simple problems first;

   d. Making complete lists of further problems.

4. *Discarding results based on causal fallacies*: The correct way to measure anything is to compare the unknown magnitude to a standard one or to check the experiment's hypothesis against a control experiment run simultaneously under controlled conditions. Comparing the energy output of a biogas plant in a given period of time with the energy output of the same plant in another period of time is the most common causal fallacy in the literature.

The following practical examples will show how to apply the methods described earlier to decide if an academic research is worth being implemented in one's own biogas plant.

6.2.1 The Correlation between BMP and Electrical Conductivity

This method was introduced in Chapter 1, Section 1.4.7., and in the present section, we will show how to rationally check its validity.

The correlation between electrical conductivity and performance of the biogas plant has been introduced by a group of Italian researchers (Garuti et al., 2014) as a possible method to "easily control" an anaerobic process. The reasoning presented by said authors is somehow reductive, although formally logic:

1. The conductivity is proportional to the concentration of dissolved salts and to the concentration of ammonium ions too (true).

2. Both concentrations have a negative influence on the metabolism of the methanogenic *Archaea* and also on the metabolism of other microorganisms involved in the digestion process (assumed true).

3. Hence, it would be possible to establish a correlation between the electrical conductivity and the methane yield of the whole process, tabulate (or plot) it, and employ such table (or curve) as control parameter for the plant.

Since its publication in a specialized agricultural magazine, many Italian biogas plant managers adopted the said method as an absolute rule, blindly measuring the electrical conductivity and logging the measures in historical series. The reason of such acceptance is the undiscussable reputation of the research institute in the Italian biogas sector, the apparent correctness of the syllogistic reasoning (two true statements leading to a conclusion), and a set of experimental values confirming the conclusion. Nevertheless, the authors state in the same article that some tests performed in real-sized plants did not yield such a linear correlation between electrical conductivity and biogas yield.

Applying the methodological skepticism leads to the following reasoning and points out what experimental evidence to search for the following:

1. *Accepting only information you know to be true*: One could argue that Archaea are organisms that have a marine origin, and hence they should be salt-tolerant. The author has personally tested sludge from marine bottom that had a good methanogenic activity. Another argument that gives rise to a doubt is that "salt" does not necessarily mean "chloride." The test was performed using a "reference saline solution" containing chloride salts of sodium, potassium, and magnesium. It is known that different salts (more specifically, different ions) have different degrees of toxicity to microorganisms. Furthermore, there are thousands of microbial species thriving together in the digester's sludge, each one having different tolerance levels to salinity. Finally, a given level of electrical conductivity may not have the same toxicity for the bacteria, depending on which the cause of said conductivity is dissolved salts or ammonia, or other salts different from chlorides. In a "normal" biogas plant's sludge, the conductivity is given by ammonia, carbonates, phosphates, and sulfates rather than chlorides. The researchers adopted such "reference saline solution" because specified in the norm UNI EN ISO 11734:2004 *Water quality—Evaluation of the "ultimate" anaerobic biodegradability of organic compounds in digested sludge—Method by measurement of the biogas production*. Such norm is applicable to wastewater treatment plants, where chloride salts are more likely to be present than in agronomical biogas plants.

2. *Breaking down these truths into smaller units*: Then the proposition (2) of the syllogism at the root of the electrical conductivity method may be not an absolute truth and if so, the conclusion derived from it cannot be absolutely true, i.e., it can be true sometimes but sometimes not. Since the "model" of salinity tested by the researchers consisted of

adding the aforementioned saline solution until the desired electrical conductivity in each set of test reactors was reached, it is necessary to prove that the same conductivity will produce the same inhibiting effect even when the salts employed are not chlorides. So the doubt is as follows: Was the electrical conductivity the real cause of the impaired methanogenic digestion observed by the researchers, or was it just the observable effect of the chlorides concentration, the toxicity being caused by the latter? Finally, as all papers in the biogas literature, the article in question does not present an analysis of the uncertainty, so one could argue that the inhibition observed may be—at least in part—just the result of instrumental or method errors.

3. *Solving the simple problems first*: The first problem to solve is whether a reactor, doped with phosphates and carbonates up to a given electrical conductivity, will show the same decrease of methanogenic potential encountered by the researchers in question. If it is so, then the proposition (2) of the syllogism is more likely to be a "universal truth," and the method of the electrical conductivity monitoring would be "universally" valid. The author prepared then a comparative test with two sets of reactors. The first set consisted of a blank reactor, a reactor with acetic acid, and another reactor with cellulose, all of them inoculated with "natural" sludge having 9.8 mS/cm and pH = 7.6. The second set of reactors was loaded with the same inoculum, to which phosphoric acid, magnesium carbonate, and potassium hydroxide were added until reaching 20 mS/cm, while keeping the pH at 7.6. At such conductivity, the resulting BMPs of the second set of reactors should be 15% lower than the control set. Since the author's instrument is an AMPTS II with 1% error margin, such a difference between both reactors sets should be easily and reliably detectable. The results are shown in Figure 6.1.

4. *Making complete lists of further problems*: According to one of the many variants of Murphy's Law, trying to solve small problems leads to bigger problems. Although such statement has no scientifically proven validity, it seems to be as inexorable as any law of physics. It can be observed in Figure 6.1 that there was indeed a reduction of the *calculated* BMP in the case of reactors with higher electrical conductivity, which seems to correspond well with the experience of the Italian research group. But we should consider all the facts affecting our experiments, and then carefully analyze Figure 6.2. It is evident that the blank reactors with the high-salinity inoculum produced more methane than the corresponding blank with the "natural" inoculum, and the same happened to the reactors with acetic acid. Such result is exactly the opposite of the expected. The shapes of the curves clearly show that the abrupt change in salinity induced by the addition of phosphoric acid and potassium hydroxide leads to some

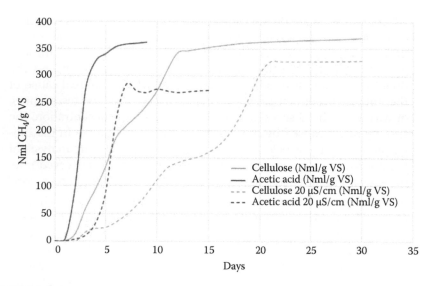

FIGURE 6.1
Specific methane production of the comparative test between two samples of sludge with different electrical conductivities and the same reference substrates.

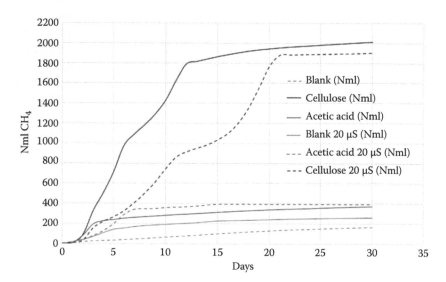

FIGURE 6.2
Gross methane production of the comparative test between two samples of sludge with different electrical conductivities and the same reference substrates.

kind of biological disruption, followed by the adaptation of the bacteria to the new conditions. The "raw" curves of the reactors loaded with acetic acid and those with cellulose end up to practically the same total methane production from the same amount of substrate, albeit with different kinetics. The higher salinity induced with P and K seems to have increased the activity of the methanogenic Archaea, and hence of the inoculum contained in the blank, and somehow reduced the activity of the hydrolytic bacteria as we can deduce from the lower production of the reactor with cellulose. Hence, the test led to an uncertain result because the differences in the *calculated* BMP were caused by a higher methane production of the *blank* and not by an impaired methane production of the *sample reactors*. There are many reasons why this may have happened, but from the biogas plant manager's perspective, our test shows that the electrical conductivity criterion is not as linear as assumed by the Italian researchers, and its practical application is not straightforward.

6.2.1.1 Conclusion

The experimental results of anaerobic digestion (AD) at 20 mS/cm electrical conductivity do not show a direct correlation between the case in which the conductivity is caused by the presence of sodium chloride and the case in which the ions are those of potassium and phosphorous. In a normal biogas plant, it is quite improbable that sodium chloride reaches relevant concentrations. In general, eventual increases in conductivity will be caused by a "cocktail" of ions: sodium, potassium, magnesium, ammonium, sulfur… In order to be valid, the electrical conductivity method should be based on a table (or plot) of values measured with the actual mix of ions that the sludge can most probably contain. Furthermore, the measures should be performed at the same operational temperature of the plant. Hence, one should employ a set of tables (or an abacus) rather than a single table or curve in order to correctly apply the electrical conductivity method. It is definitely much more reliable to measure the BMP of reference substrates at periodic intervals with own inoculum of the digester, in order to check its *real* microbial activity, than trying to deduce it from an indirect measure.

A final reflection for readers belonging to the academic world: when researching on the correlation between the salinity of the sludge and its microbial activity, the mineral content of the feedstock should be taken into account. For instance, biogas plants fed with corn silage (most European ones), equipped with sludge recirculation, will be more prone to accumulate salts with time than plants without sludge recirculation, in which the dilution of the feedstock is mainly done with water. A useful research on the correlation between the electrical conductivity and the methanogenic activity of the sludge should be performed using a saline solution, which is really representative of the kind of salts found in the sludge itself. For instance, according to a study of the Food and Agricultural Organization (FAO), the mineral fractions contained in corn biomass are the ones shown in Table 6.1. Hence, the saline solution added to the sludge to increase its electrical conductivity should contain the same minerals, and in the same proportions listed in the said table.

TABLE 6.1

Mineral Content of Corn According to
FAO (Average of Five Samples)

Mineral	Concentration (mg/100 g DM)
P	299.6±57.8
K	324.8±33.9
Ca	48.3±12.3
Mg	107.9±9.4
Na	59.2±4.1
Fe	4.8±1.9
Cu	1.3±0.2
Mn	1.0±0.2
Zn	4.6±1.2

The salts should be mainly carbonates, phosphates, and sulfates, which are the most likely species found in agricultural sludge. Sodium should be present both as bicarbonate and chloride. A more pragmatic approach could be calcining a certain quantity of biomass in the muffle oven and then using the ash to prepare the saline solution for the tests. Such saline solution will be surely more representative of the specific digester's operational conditions than the standard saline solution defined by norm UNI EN ISO 11734:2004 that was adopted for the basic research quoted at the beginning of this section.

6.2.2 Assessing the Validity of the VFA/TA (FOS/TAC) Method

This method was introduced in Chapter 1, Section 1.6. Its wide diffusion can be attributed to the marketing efforts of the company that produces the most popular *ad hoc* automatic titrator and the sales argument, "Thousands of biogas plants in Germany use it." The limited validity of the said method can be proved by applying the simple rules of the syllogism:

1. The volatile fatty acid (VFA) concentration is an indicator of the stability of the AD process (true, but only under certain conditions).
2. The alkalinity is an indicator of the resistance to changes in pH (true).
3. Hence, by correlating the VFA concentration and the alkalinity, for instance, by means of a simple ratio, we can obtain a measure of the stability margin of the AD process [only when (1) is true, but under certain conditions it can be a misleading conclusion].

The experimental evidence already presented in Section 1.6 should be enough to prove the Cartesian doubt: The FOS/TAC (*Flüchtige Organische Säure/Totales Anorganisches Carbonat*) cannot be universally applied because it is not based on universally valid suppositions. From a practical point of view, a method that tells us that everything is running properly when we already know that there is no problem—but in many cases does not provide any useful information to diagnose why the anaerobic process is blocked— is unreliable. By applying the methodological skepticism, the reader could argue that the study case presented in Section 1.6 is just one example of a biogas plant that for some reason suffered a severe inhibition of the Archaea. Hence, the following experiment will dissipate any doubt:

1. Take some digestate from a "healthy" biogas plant and sieve it with a 5-mm mesh (a kitchen sieve like the one employed for spaghetti will do).
2. Fill two reactors with equal amount of digestate and incubate it for at least 5 days.
3. Add 2 g of acetic acid per liter of inoculum to one of them.
4. Add 2 g of propionic acid to the other.
5. Measure the VFA/TA (total alkalinity) of both if you have any doubt. They should be approximately the same, since the VFA/TA method does not make any difference between different VFA.
6. Incubate the reactors for at least 4 days and check how much methane each one produced.
7. Convinced now?

6.2.3 The Causal Fallacy

A logical fallacy is a false reasoning that seems to make sense. The most common type is the causal fallacy, in which the cause of an observable effect is incorrectly identified. Classical logics call it the *post hoc* fallacy because the Latin sentence describing it as *post hoc, ergo propter hoc* ("after this, therefore because of this"). This is the kind of sales argument usually employed by producers of "boosters," additives, or machinery that allegedly increase the efficiency of an anaerobic process. From a scientific point of view, the differences observed in the yield of a biogas plant before and after applying a certain product can have tens of causes, which are not related directly to the single cause assumed as the source of the differences observed before and after. The following example shows how to correctly evaluate the efficacy of enzymes, promoted by some articles on the base of "before" and "after" tests in full-sized biogas plants.

6.2.3.1 Information from the Literature

1. According to M. Plöchl et al., applying enzymes reduces the viscosity of the sludge, though the effect is small. The paper does not include any consideration on the biogas yield, only the advantages of reduced viscosity, i.e., less self-consumption of energy in the stirring system.

2. According to Bruni and Agelidaki, treating manure and straw with OCa increased the biogas yield by 66%, while steam treatment followed by enzymatic treatment only gave 35% increase.

3. According to the chemical company Dupont, applying their enzymes increases the biogas production by 15%; but in another paragraph of the same brochure, it is stated that the net methane production increases only 8%.

4. According to Petta et al., the addition of enzymes to a biogas plant enabled saving 8% of the feedstock for the same energy production.

6.2.3.2 Checking the Logic Flaws

Papers (1) and (2) contain research performed under laboratory conditions, i.e., comparing control samples with enzyme-treated samples, though no estimation of the uncertainty is included. In spite of being scientifically sound, at least from the formal method, the study of Bruni and Angelidaki has little applicability to real-size plant operation. Since they tested the enzymatic treatment after a steam treatment, it is not clear if the increase of the biogas production is caused mainly by the first or by the second. On the other hand, publications (3) and (4) are not scientific papers published in peer-reviewed journals and contain "field results" measured directly on real-size plants, "before" and "after," i.e., the method employed arises the suspicion of causal fallacy. In both cases, the enzymes employed are extracted from fungi, i.e., they are mixtures of cellulase and hemicellulase, probably also some laccase. In both cases, the reported improvement is the average of a high number of individual measures over time, but there is no description of the instruments employed for taking said measures and their uncertainty range.

6.2.3.3 Checking the Efficacy of Enzymes in the Correct Way

The author performed a simple laboratory test by simply taking sludge from the primary digester of a biogas plant, which contained 9% volatile solids (VS) and performing a comparison between the methane yield of the control reactor and the reactors to which enzymes were added. The sludge contained mainly vegetal biomass and cow manure, hence two different commercial enzymes were tested: one is cellulase in liquid form (300 mg/L recommended dosage) and the other is a mixture of cellulase and hemicellulase (recommended dose 30 mg/L). Figure 6.3 shows the result after 30 days digestion at 38°C. Observe that the curves of the methane production with enzyme addition are almost coincident, showing that both commercial products are functionally equivalent. The increase in net methane yield was 9%. Since the error margin of the instrument is 1%, we can conclude that the application of such enzymes leads to improvements comprised in the 8%–10% range.

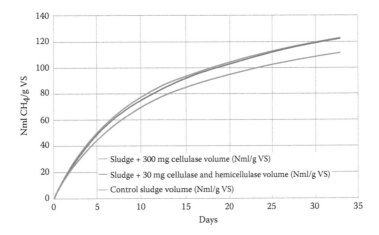

FIGURE 6.3
Comparative AD test of sludge containing high amounts of vegetal fiber, with and without enzyme addition.

An important consideration, not specified in any of the cited publications, is that the activity of enzymes depends on their concentration. This means that the dosage must be calculated in two steps: the initial amount for the first time, as a function of the useful volume of digestion, and the daily dose, which depends on the volume of feedstock and water added each day, useful to keep such concentration constant in time. Example: Suppose a digester having $3000\,m^3$ of useful volume. The same is fed with $100\,m^3$ (solid feedstock plus dilution water) everyday. The enzymes dosage recommended by the manufacturer is $30\,mg/L = 30\,g/m^3$. Hence, if the plant manager decides to employ such an enzyme, the operation strategy must consider an initial load on the first day equals to $90\,kg$—in order to bring the whole digester to have an adequate enzyme concentration—and $3\,kg$ everyday, in order to keep such concentration constant.

In this particular case, the results measured in laboratory conditions are in line with those deduced with the "before" and "after" experiments.

6.3 Misconceptions of the Scientific Literature Amplified by the Marketing

6.3.1 The Importance of pH and the Use of Sodium Bicarbonate as Buffer Agent in Anaerobic Plants

Adding sodium bicarbonate to the digester, with the scope of "keeping a stable FOS/TAC value" or "avoiding acidification," is a relatively common practice in some European countries. Adding sodium bicarbonate serves only to mask eventual biological problems in the digester, without eliminating their causes. Furthermore, the indiscriminate use of sodium bicarbonate, without laboratory tests determining its effective need, is nothing but a useless cost and a potential danger for the fertility of the soils that will later receive the digestate.

In principle, it is true that adding sodium bicarbonate helps maximizing the production of acetic acid, the main nutrient of the acetoclastic Archaea, but it is also necessary to consider the following matters:

1. Sodium bicarbonate is not the only cheap product for neutralizing the acidity: lime, potash, and magnesium hydroxide are also useful. The said alkalis are the main constituents of biomass ash.

2. Agricultural biogas plants are often fed, in variable proportions, with three substrates that feature high alkalinity: bovine manure, swine slurry, and chicken or hen dung. In Germany and Italy, instead, a consistent number of plants are fed exclusively with corn silage, cereal meals, and similar easily fermentable feedstock, with little or no manure at all in the mix, hence having scarce alkalinity. It is then natural that in such contexts, it becomes necessary to artificially provide some alkalinity in order to compensate the VFA produced by the fermentation. Such a management method should be considered as exceptional, applicable only to biogas plants fed exclusively with corn silage or saccharine or starchy substrates and not as a rule for all biogas plants.

3. Adding sodium bicarbonate to a digester that shows some tendency to acidity may cause an unwanted phenomenon: foaming. This is just the result of a chemical reaction and has nothing to do with the biological activity.

Figure 6.4 shows an experimental evidence of the former considerations. The author measured the BMP of a very acid substrate, cheese whey. The inoculum was sampled from a "normal" agricultural biogas plant, usually fed with 10% bovine manure and 90% with mixed cereal silage, all the said mixture being diluted with swine slurry. A reactor was fed with pure whey, pH = 4.9, and another reactor was fed with the same whey that had been previously neutralized with sodium bicarbonate (pH = 7). We can observe that the difference between both curves is irrelevant, lying within the typical error margin of the BMP assay.

If the usefulness of sodium bicarbonate is so limited, why is it then so diffused in the biogas industry? Most probably because the technical literature distributed by one of the multinationals promoting it claims the efficacy of its product based on "studies carried out at the University of Rostock." The same technical sheet does not quote the title, nor the author, nor the journal where such studies were published. One of the said studies, published

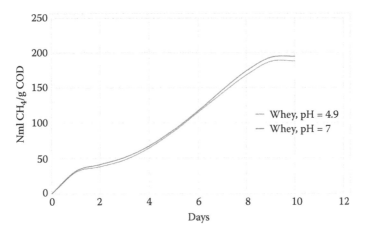

FIGURE 6.4
Comparison between the anaerobic digestion of "raw" cheese whey and the same neutralized with sodium bicarbonate up to pH = 7. (Error margin of this single test = ±5%).

in German with only the abstract in English, can be downloaded from the Internet (see Bibliography at the end of this chapter). From a careful reading, the following aspects appear questionable:

1. The methodology adopted is not rigorous because the tests were performed directly in two 550 kW biogas plants, not in a laboratory.
2. The merit parameter adopted to judge the validity of sodium bicarbonate as additive is the FOS/TAC ratio (furthermore not even measured by the researchers themselves but measured by the plant's manager).
3. As usual, the paper contains no analysis of the measure errors, which are presumably high because the measures were taken with instruments homologated for industrial scopes.
4. The text states that the systematic addition of sodium bicarbonate increases the production of methane, but the tables with the results show only the gross biogas production of each period, while the net production of methane is displayed in a graphic, with a very small scale, and has not been integrated in each period;
5. The BMP of corn silage is expressed in terms of dry matter and not in terms of VS; hence it is not possible to check if the digester's organic load (OL) was really constant throughout the test period.
6. The difference between the gross biogas production of both plants—a control plant and a test plant to which sodium bicarbonate was added—results in the same order of magnitude of the (estimated) measure error. Furthermore, stating the production of biogas instead

of the net methane production results in the following legitimate Cartesian doubt: Was the methane content the same in both plants all the time? The paper does not explain how to fairly compare two biogas plants based on the "gross biogas productions," since we all know that the methane content is variable along time, hence assuming 60% methane all the time is a mistake. Figure 6.5 shows that, after a "shock treatment" with sodium bicarbonate, the overall productivity falls drastically and is comparable to the one of the control plants only from the 60th day of the test. After 100 days of the test, the plant to which the bicarbonate was added seems to have performed better than the control. At this point, we may ask ourselves if such a result is really the merit of the bicarbonate addition, or is just a variability resulting from the combined uncertainties introduced by differences in the management of the corn silage, measure errors of the load weighting systems, starts and stops of the generators, and, above all, the error arising from the method adopted by the researchers to estimate gas production. For sure, most agricultural biogas plants in Europe usually do not include gas volume meters. In the case of the paper under analysis, it is not clear how the gas production was measured. From a very synthetic description, it seems that the researchers calculated the biogas flow with a "rule of thumb" based on the inside pressure of the gas collection bag,

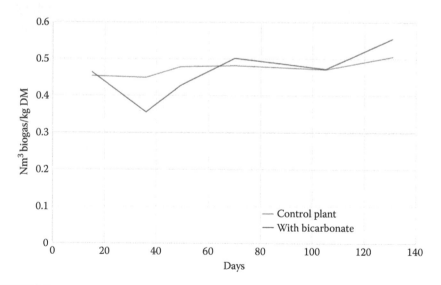

FIGURE 6.5

Gross biogas production measured in two different biogas plants: one had suffered from acidification and was subject to a "shock treatment" with sodium bicarbonate, followed by periodical additions; the control plant received no additive at all. (The graphic was plotted by the author based on the tables published in the quoted paper.)

assuming constant volume. It is evident to anybody on just common sense considerations that flexible biogas holders cannot be considered perfect hemispheres since their volume will vary enormously not only with the inner pressure, but also with the difference of temperature between inside and outside, with the solar radiation, wind pressure, snow load, etc. Should anybody have any doubt: Both the norm VDI 4630 and the draft of Italian norm on BMP measuring state that barometric measure systems must be mounted on *rigid* reactors. Hence, the validity of the conclusions drawn by the authors of said paper is at least dubious.

Another aspect that the biogas plant manager cannot ignore is the accumulation of sodium in the digestate. For instance, the dosage of sodium bicarbonate recommended by the manufacturer for a biogas plant with 1 MW electrical power—in the range of 100–200 kg/day—can in part inhibit the Archaea and end up reducing the production of methane. If the dosage mentioned before is kept constant, within 1 year, the total amount of sodium bicarbonate cumulated in the digestate will be in the range of 36.5–73 ton, with the risk of salinizing the soils instead of fertilizing them. Hence, before deciding to systematically add sodium bicarbonate to the digesters, it is quite advisable to ask an independent agronomist to check if the cumulated concentration of sodium can be dangerous for the soil's fertility. The study on soil salinization performed by the FAO and quoted in the Bibliography can serve the reader as a guideline for such verifications.

6.3.2 The Hyped Importance of Trace Elements

As in the case of sodium bicarbonate, industries, "biologists," and "gurus" of the AD merrily prescribe the addition of tons of "integrators" and "special products." The sales argument sounds always as an ominous warning: "If you do not add these products, your plant will lose productivity and risks the biological collapse." We have already explained in Chapter 1, Section 1.1.2.6 that it is necessary to supply the Archaea with *very small quantities* of some minerals called "micronutrients," "trace elements," "trace minerals," or "oligoelements" depending on the different bibliographic sources. The said small doses of minerals are necessary as catalysts for the enzymatic processes involved in the anaerobic digestion. It is very easy to find in Internet some "recipes" of mineral mixtures studied by different researchers. Nevertheless, the dosages recommended by the different studies vary enormously (see, for instance, Lebuhn et al., 2008). Not surprisingly, companies selling such mixtures always recommend the highest dosages found in the literature. Some biogas plant managers tend to exceed in the addition of trace minerals, fearing that the productivity of the digester may fall if they do not keep high concentrations of them. The truth is that not all trace elements may be necessary, and furthermore the highest dosage is not always the best solution.

According to Gustavsson (2012) and Yee Yaw Choong et al. (2016), the aceto-clastic Archaea group requires the availability of trace elements, in particular, Ni, Co, and Fe, while the hydrogenothrophic Archaea are inhibited if the concentration of such minerals exceeds a given threshold. It is then clear that the dosage of minerals must respect a very delicate balance—too high will inhibit the hydrogenothrophic Archaea, hence the proportion of CO_2 in the biogas will rise. Furthermore, the mere presence of trace elements in the sludge is not enough to grant a good methanogenic activity—the minerals must be present in a form that is soluble in water and that the Archaea should be able to assimilate (e.g., sulfates, carbonates, phosphates,...). As a further complication of the problem, the concentration of SH_2 plays a role in the assimilability of trace minerals, since SH_2 may react with some metallic salts, precipitating them as insoluble sulfides. The presence of phytic acid, usually in high concentration in those biogas plants fed with cereal silage and very little or no cow manure at all, affects negatively the balance of trace elements. Phytic acid tends to chelate metallic ions, making them unavailable to the Archaea. Ruminant's manure contains phytase, an enzyme that breaks the chelates. So, the amount of trace elements that should be added to the digester in order to keep its activity at optimum level is also a function of the percentage of cow manure in the digester's diet.

The easiest way to determine whether a plant needs the addition of trace elements, and how much of them, is to perform several specific methanogenic activity (SMA) tests on the same inoculum, each one with a different concentration of minerals. Figure 6.6 is an example of such a test. In this particular example, the inoculum had already a sufficient concentration of trace elements, so the addition of "medium" (micronutrients solution) led to a *reduction* of the SMA instead of the expected increase.

6.3.3 The Use of "Special Products" for Desulfurization

All biomasses contain sulfur in different concentrations. The degradation of organic matter by the metabolic activity of the anaerobic microorganisms releases sulfide acid, SH_2, which is an inhibitor of the biological activity and corrosive for metallic parts. It is then necessary to desulfurize the biogas, mainly for protecting the motor. The existing desulfurization technologies can be either biological—carried out by bacteria that precipitate the SH_2 as free S—or chemical, neutralizing the acid with alkalis (lime, caustic soda, sodium carbonate, or bicarbonate) or ferric compounds (especially those containing iron in trivalent state). Most European biogas plant constructors do not include dedicated (i.e., external) desulfurizers, as it is the case in the USA, China, and India. On the contrary, the European biogas industry usually prefers the *in situ* desulfurization (i.e., in the digester itself). Such technique requires injecting a small quantity of air in the digester, necessary for the sulfur-reducing bacteria, but causing anyway a small reduction of the methane concentration in the biogas. Such method is economical (for the biogas

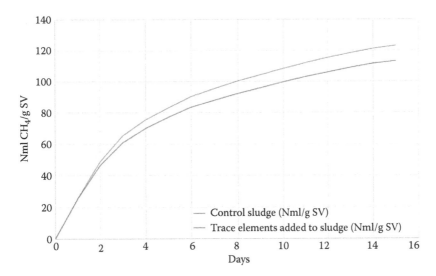

FIGURE 6.6
Effect of adding trace elements to an inoculum that does not need them. The result was 8% less methane in 15 days, compared to the control reactor. The author made this test with inoculum sampled from a plant usually fed with silage, vegetables, olive mill pomace, and cow manure that showed no sign of problems. The test was performed with an AMPTS II, and perfectly preincubated inoculum, resulting in an overall error margin of 1%.

plant constructor!) but not always fully effective in eliminating sulfur; hence in some cases, the plant manufacturer requires or specifies adding some "special products," generally the most expensive among the possible alternatives and for sure, always more expensive than their equivalent commodities available in the industrial chemistry market.

The cheapest and most environmentally compatible desulfurizer, which no manufacturer or "specialist" wants to recommend, is quicklime (calcium oxide, OCa), just because it is easily available in the market and allows the biogas plant owner to be independent. Quicklime reacts with the dilution water, forming spent lime (calcium hydroxide, $Ca(OH)_2$). This one can either react directly with the SH_2 or with the CO_2 dissolved in the sludge, forming, respectively, chalk (calcium sulfate, SO_4Ca) or calcium carbonate (limestone, CO_3Ca). Calcium carbonate, in turn, reacts with SH_2 and precipitates as chalk. The drawback of using lime as desulfurizer is the sediment that can grow with time if the stirring of the digester is not strong enough. The presence of chalk and limestone does not alter the fertilizing properties of the digestate, since both compounds are usually present in soils.

Some companies promote the use of sodium bicarbonate as desulfurizer. We already discussed the use of bicarbonate as buffer in Section 6.2.1. Using it as desulfurizer is analogous to using lime, with the difference that sodium does not precipitate but its presence in the digestate may reduce the fertilizing properties of the latter. The desulfurizers preferred by constructors and companies that manage biogas plants for third parties, and at the same time sell such products, are iron compounds. From a technical point of view, such choice is correct because trivalent iron—present in ferric oxide (O_3Fe_2), ferric hydroxide ($Fe(OH)_3$), and ferric chloride (Cl_3Fe)—is very effective in selectively neutralizing SH_2. Iron compounds combine with sulfur, precipitating as insoluble salts, in the ways described by the following reactions:

1. Iron(III) oxide (ferric oxide)

$$3SH_2 + O_3Fe_2 \text{ (in aqueous medium)} \rightarrow S_3Fe_2 + 3H_2O$$

2. Iron(III) hydroxide (ferric hydroxide)

$$2Fe(OH)_3 + H_2S \rightarrow 2Fe(OH)_2 + S + 2H_2O$$

3. Ferric chloride

$$3SH_2 + 2Cl_3Fe \rightarrow S_3Fe_2 + 6HCl$$

4. Ferrous hydroxide (bivalent iron)

$$Fe(OH)_2 + H_2S \rightarrow FeS + 2H_2O$$

From a commercial point of view, the said technical choice generates important incomes for those companies selling "special desulfurizers" to biogas plant managers, who in good faith believe they are buying the best products to protect their motors from corrosion. Table 6.2 shows a comparison between three offers from different suppliers: an Italian constructor of biogas plants, a German laboratory service company that sells "special" products for biogas plants, and a wholesaler of standard commodities for the chemical industry.

TABLE 6.2

Comparison between Three Commercial Desulfurizer Products

Supplier	Product	Composition, According to Technical Sheet	Effective Fe^{+3} Content	Impurities	Price (Italy, Sept. 2016)
Italian constructor of biogas plants	Fantasy name X	$Fe(OH)_3$	36%–40%	Pb, Cd, Cr^{+6}, Cu, Ni, Hg, Tl, Zn	1.20 €/kg
Biogas-specialized German services company	Fantasy name Y	$Fe(OH)_3$	38%–41%	Pb, As, Cd, Ni, Cu, Cr, Zn	1.42 €/kg
Wholesaler of industrial chemical commodities	Ferric oxide (CAS N. 1332-37-2)	O_3Fe_2	66%	Al_2O_3, SiO_2, MnO	0.87 €/kg

The first two have named their "special desulfurizers" with fantasy names, quoted as X and Y in this text for copyright reasons, but according to their technical safety sheets both products are just iron(III) hydroxide. By the way, not precisely of the best quality…

The former example shows how scientifically sound information from the literature is being exploited to sell low-quality products at higher prices than the market average for the same commodity. It also shows how the European biogas industry saves on installation costs by choosing the *in situ* biological desulfurization and not installing external desulfurization units, charging the operational cost of iron compounds to the biogas plant owner—and earning money on top—when the biological system is not able to bring the SH_2 content in the biogas below the admissible limits.

6.3.4 Databases and Mathematical Models of BMP

Some European research groups and constructors of biogas plants promote large databases of BMP of many substrates as fundamental tool for designing and managing biogas plants. There is a generalized belief, even in academic circles, that such databases are enough to predict the methane yield of any substrate under any condition, as if the BMP was a defined and invariable property of matter. The European Biogas Association (http://european-biogas.eu/) has collected the links to seven different Internet sites offering the possibility of free access to such databases. Table 6.3 shows a comparison between the BMP of some usual biomasses taken from three of such databases. The reader can draw his own conclusions about the usefulness of such data.

Should there be any doubt, the following experimental data will dissipate it: A sample of microcrystalline cellulose—a pure substrate with a well-defined theoretical BMP—was tested with an AMPTS II using two different

TABLE 6.3

Comparison between the BMPs of Some Common Agricultural Substrates for Anaerobic Digestion, According to Different Sources

Substrate	BMP According to Different Source [Nm³/ton VS]		
	KTBL (Germany)	LFL (Germany)	Biowatts (Germany)
Corn silage	310–411	294–372	305–326
Triticale silage	281	259	259–606
Grass silage	318	295–329	315–330
Cattle manure	209	154–247	122–220
Pig manure	336–453	240	240–273
Poultry manure	293	325	275
Meat waste	583	—	733
Glycerol	751	425	425
Horse manure	—	165	165

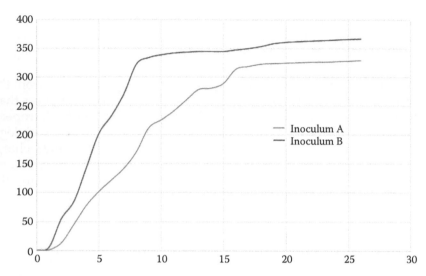

FIGURE 6.7
Anaerobic digestion of a single sample of pure microcrystalline cellulose with two different inoculums.

inoculums, sampled from two biogas plants running normally. The test was run in duplicates, with an error margin of 2%. Figure 6.7 shows the results: The difference between the methane yield of both inoculums was 11%. Hint to solve the riddle: inoculum A was sampled from a digester treating wastewater sludge, while inoculum B was sampled from an agricultural biogas plant, usually fed with bovine and swine manure and cereal silage.

6.3.4.1 Conclusion

If only the quality of the inoculum can induce 11% variation on the measured BMP, not considering any other experimental errors, then: What is the reliability of tabulated values, not knowing the kind of inoculum, the method, and the error margin of the test from where the said values come from?

6.3.5 The Conservation of the Inoculum

Norms on biological tests, as for instance, the VDI 4630/2014, usually state that the inoculum must be sampled and conserved at 4°C until the moment of the test, to prevent the bacteria from dying and affecting negatively the test result. It is known to all biologists that bacteria and Archaea are very resistant organisms. Some papers in the literature state that the inoculum for BMP was sampled and kept at room temperature before starting the test (e.g., Luna del Risco et al., 2011). Hence, the Cartesian doubt is fully legitimate: Is it really necessary to keep the inoculum refrigerated until the start of the test? It is highly improbable that a biogas plant manager needs to conserve the inoculum for long periods of time, since he has the possibility of sampling it when necessary. The present paragraph was included in this book for the sake of researchers and also biological assistance professionals, who perform routine tests for commercial biogas plants. Both categories of laboratory operators may face the dilemma of sampling inoculum from the plant each time they need to carry out a test or keeping in the lab's fridge a certain quantity of inoculum. In both cases, they will have to bear a cost either for the displacement to the plant or for the space and energy required by the refrigerator(s). The author decided to perform a very simple experiment to check if long periods of storage at uncontrolled temperature can hamper the inoculum's bacterial activity. Two routine tests with cellulose and acetate were performed with fresh inoculum. Another sample of the same inoculum was kept in a hermetically closed plastic container, placed in a patio from November 2015 to November 2016, vented from time to time to avoid gas overpressure. The same routine tests with cellulose and acetate were then performed in December 2016. The results are shown in Figure 6.8.

The following can be observed:

1. The aged inoculum has a longer lag phase. This is because after 1 year at varying temperatures, the microorganisms needed some extra time to adapt to the mesophilic conditions of the test.

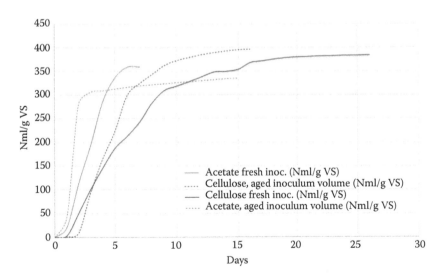

FIGURE 6.8
Hydrolytic and methanogenic activity of freshly sampled inoculum, compared to the same inoculum after 1 year aging at variable outdoors temperature.

2. After the lag phase, the aged inoculum showed a higher kinetic constant (steeper curves) than the one measured when it was fresh. This is because the microorganisms had depleted all the available substrate during the aging period and were starving.

3. The final BMP values of acetate and cellulose obtained with fresh and aged inoculum are not directly correlated to the age. One should have expected aged inoculum to yield lower BMP, which was indeed the case with acetate but the opposite with cellulose. Anyway, the activity of the aged inoculum was still acceptable.

The extreme test described here shows that even keeping the inoculum during one full year, at temperatures varying from –2°C to 33°C (Northern Italy), has not affected too much the hydrolytic and methanogenic activity. We can safely assume that keeping the inoculum indoors at room temperature for a few months will not compromise the results of the tests.

6.3.6 The "Equivalent Corn Silage Unit"

Most European biogas plants were designed to run on "energy crops," usually corn silage. Such questionable design choice is the result of incentive policies adopted by the European Union on the base of a business model promoted by several key sectors of the German industry: the mechanical, electrical, chemical, and seeds industries together with banks and investment groups. The reasoning supporting the use of corn silage as feedstock for producing energy is an example of simplistic linear thinking: a single substrate—assumed as having a constant methane yield—a table with some standard parameters to control the process, a generator running 99% of its useful life at maximum power (unrealistic, but investors like it), hence return of investment fully warranted by State subsidies. The pressure of environmentalist groups and many studies shows that cultivating corn is not a sustainable way for producing energy, led to redimensioning the subsiding policies. For instance, current (2016) regulations in Italy limit corn silage to <30% of the total amount of "energy crops" in the biogas plant's feedstock. As a natural consequence, biogas plant managers began then sourcing other substrates. Since all their previous calculations were based on the yield of corn silage, this latter became a new measure unit to calculate the digester's diet: the *equivalent corn silage unit (ECSU)*.The *ECSU* is defined as the quotient between the methane yield of one ton of feedstock X and one ton of corn silage (assumed as "universal constant," equal to 115 Nm³/ton w.w.). The simple application of the syllogism method already demonstrates that the *ECSU* is not a reliable criterion:

1. The «standard» BMP of corn silage is 115 Nm³/ton w.w. (FALSE! The BMP of corn silage is highly variable).

2. The BMP of a given biomass is X Nm^3/ton w.w. (maybe true in a given moment, but strongly dependent on the moisture, which is variable in time).

3. Hence, each ton of feedstock X can replace X/115 tons of corn silage (may be true just for pure case, but in principle this conclusion is FALSE).

The falsity of the first proposition was already demonstrated in Section 6.3.4 and with the example shown in Figure 6.7. The effective methane yield of a substrate in a given biogas plant in a given moment has nothing to do with the theoretical BMP; it is a combination of several factors, the inoculum's bacterial biodiversity being one of the most important ones.

The *ECSU* criterion may work if another silage is intended as replacement for corn silage. Even if it is absolutely false that the BMP of corn silage is constant and equal to 115 Nm^3/ton w.w., the said value is quite frequent. If the physicochemical features of "feedstock X" are similar to those of corn silage (e.g., similar density and moisture content), then the *ECSU* criterion may work, even in plants having some biological problem. The reason is that the digestion conditions in the biogas plant being the same for corn silage and "feedstock X," then the nondimensional ratio defined by the *ECSU* remains constant, independently of the effective methane yields of both substrates being lower than those obtained in the laboratory. Again, this is one of those cases in which the syllogism turns to be true but just by pure chance.

The following examples show the most common cases in which the concept of *ECSU* is unreliable for a correct formulation of the digester's diet.

6.3.6.1 Olive Mill Pomace

The BMP of this substrate varies not only as a function of the digester's bacterial ecosystem but also with the kind of olives, their degree of maturation, the oil extraction process, etc. Furthermore, olive pomace requires long digestion times of 45–60 days, while corn silage exhausts its methane potential within 15–20 days.

Hence, even if the "tabulated BMP" is nearly the same for both substrates (*ECSU* ≈ 1), replacing a given daily quantity of corn silage with the same quantity of olive pomace can pose a serious risk of inhibiting the bacterial activity.

6.3.6.2 Fatty Waste and Glycerol

Such substrates have very high theoretical BMP; hence ECSU >1, i.e., a small quantity of such substrates can replace large quantities of corn silage, an appealing possibility for biogas plant managers who need to buy biomass for their digesters. Nevertheless, the BMP of such substrates is a function of the OL (i.e., the I/S ratio), and the experience shows that they can be strong

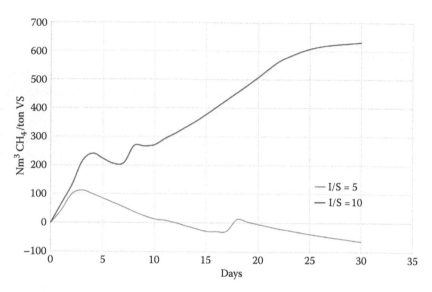

FIGURE 6.9
Inhibition caused by glycerol even at high I/S ratio.

inhibitors. Hence, replacing corn silage with its ECSU of glycerol of fatty waste will most probably inhibit the bacterial activity, eventually stopping or strongly reducing the energy production and its associated income. By pure chance, some biogas plants may not encounter any problem at all (e.g., digesters inoculated with sludge from sewage treatment plants, or plants fed regularly with fatty waste). From the author's experience, such plants are to be considered as exceptional cases, so the positive results observed in them cannot be generalized to all biogas plants. Figure 6.9 shows the digestion of the same sample of glycerol at different I/S ratios. The inoculum was sludge samples from a biogas plant usually fed with corn silage.

6.3.6.3 Mixtures of Substrates

In the world of anaerobic digestion, 2+2 sometimes may be equal to 4, but in some cases, it can yield 5 or just 3. The mixture of two or more substrates is the typical case when the total *BMP* is not proportional to the individual *BMPs* of the single components. Figure 6.10 shows the result of laboratory tests with mixtures of bovine and rabbit manure. The *BMP* of the mixture depends strongly on the proportion of both substrates, resulting in maximum when the VS of rabbit manure are four times those of cow manure (which in this test was stall manure, mixed with straw). Hence, should we try to apply the *ECSU* criterion to calculate a diet based on the mixture of said substrates, the theoretical result would be very different from the reality. The reason of such apparent paradox was already explained in Chapter 1: one of the parameters that influence the value of the *BMP* is the *C/N ratio*.

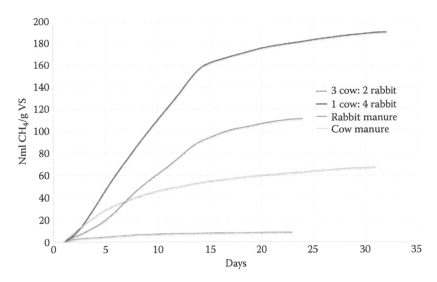

FIGURE 6.10
Mixtures with different proportions of the same substrates will yield different total BMPs.

6.4.6.4 Substrates with Low Concentration of Degradable Organic Matter

Another limitation to the use of the ECSU criterion for replacing silage with an "equivalent" substrate is the useful volume of the digester. For example, consider the AD of bovine slurry shown in Figure 6.11.

Such a substrate has 96% water and its BMP is equal to 0.023 *ECSU* (i.e., 42.8 tons of slurry are needed in order to produce the same quantity of methane as 1 ton of corn silage).

A biogas plant consuming 50 tons/day of corn silage (approximately 1 MW power) would then require 2141 tons/day of such slurry, which means occupying approximately 2141 m³. Observing Figure 6.11, we notice that the sample required 30 days for its complete degradation. Considering that the useful volume of average $1 MW_{el}$ plants (in Europe) is about 6,000 m³, the impossibility of applying the *ECSU* criterion in this case is quite evident: the digester's HRT would be just 3 days, which is not enough to digest the slurry.

6.3.7 Estimating the Accuracy and Reliability of Data Published in the Literature

Quite often the biogas plant manager needs to include a new substrate in the diet of his digester, and faces the old dilemma: "Which is the BMP of this feedstock?" We already explained that collections of BMP values are not reliable, so one could be tempted to look for answers in specific papers, where the authors and the test methods are known.

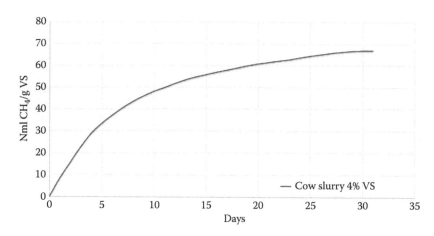

FIGURE 6.11
Anaerobic degradation of bovine slurry having 4% VS.

In spite of the existence of specific norms[*] on assessing the uncertainty margin of laboratory measures, a constant in all the literature is the lack of such kind of analysis. Most of the papers on BMP of different substrates provide tables reporting the average values of several measures and their dispersion. On one side, such values sometimes represent the "raw" gas production, which is unrealistic because it includes the moisture. From a metrological point of view, checking only the standard deviation of a set of measures is not fully correct, since it is an indicator of the precision (i.e., repeatability) of the test, but not of its accuracy. The accuracy is especially important when the measured values must be employed for further elaboration—e.g., the energy value of the feedstock as a function of the gas production—since calculations amplify the error of the final result, according to the rules and theorems already explained in Chapter 2. Furthermore, gases vary their volume with temperature and pressure as well as their moisture. In many cases, the value presented is the total biogas production and the methane content, this latter being the result of a separate measure that has its own error. Finally, the quality of a laboratory measure can be determined only by comparing the uncertainty of the average value with the dispersion or standard deviation of the replicate measures. If the error is of the same order of magnitude of the standard deviation, then the average value can be considered at the same time accurate and precise (the best quality of test with a given set of instruments). Should the error be bigger than the standard deviation, then the test can be considered precise (i.e., correctly performed) but not accurate (i.e., the average value is not reliable because the instruments and methods amplify the measure's uncertainty). In such a case, the lowest limit should be adopted as a conservative value for any calculation. Should the error be

---

[*] The following are some of the pertinent norms:

smaller than the standard deviation, then the average value is accurate and can be employed for calculations, even though the test is not precise (e.g., because the substrate is heterogeneous).

The following example shows how to analyze data from the peer-reviewed literature, in order to decide if they are reliable and useful for practical scopes.

The paper taken as example is *Investigation of the methane potential of horse manure* by Mönch-Teder et al. (complete reference and link for free download in the Bibliography). According to the authors of the said study, horse manure has a very variable BMP for two reasons: the diet and metabolic state of the animal influencing the composition of its dung, and the assortment of bedding materials employed by different stalls, ranging from almost indigestible wood sawdust to fairly digestible wheat straw. The said study is very detailed, since dung from many stalls and with different beddings was tested and analyzed. We will check one single value: horse manure from stall A containing straw, and then analyze the remaining values from stalls B, C, D, and E, which contained straw too. The test protocol employed 30 g inoculum and 600 mg dried and pulverized manure, tested by triplicates with a blank. The BMP resulting from the normalization and average of the single samples from stall A was 0.176 Nm³/kg VS±1.73%. Note that 1.73% is the dispersion coefficient of the three replicates (i.e., the standard deviation expressed in % of the average), not the error margin of the measure. The procedure to estimate this latter is based on the following reasoning:

Graduated laboratory glassware can be labeled as class A (error = 0.5%) or class B (error = 1%), according to ISO 4788 (equivalent to DIN 12600). The paper states that the 100-ml syringes have 1-ml graduations, so we deduce they are class B and their absolute error is hence ±1 ml. Since they were filled with 30 ml of slurry and 600 mg substrate, the total volume of methane produced was:

$$V_{methane} = 0.6 \text{ g} \times 176 \text{ Nml/g VS} = 105.6 \text{ Nml}$$

Since the test temperature was 37°C, the volume of methane at this temperature was as follows:

$$V_{methane} = 105.6 \times \frac{310 \text{ K}}{273 \text{ K}} = 120 \text{ ml}$$

The paper does not state the composition of the biogas of each single measure. For simplicity, we can estimate that it contained 60% methane, so the total volume of biogas produced was in the order of 200 ml. Since the syringe only has 100 ml total capacity, and was filled with ~30 ml, the maximum admissible piston excursion is 70 ml. This means that at least one reading must be taken before reaching said point. If we observe the curves included in the paper, they were traced with 18 points. We can then deduce that each reading interval corresponded to an average of 11 ml of biogas production.

The total error of the sum of 18 individual measures can be then calculated with the rules of error propagation:

$$e_{biogas} = \frac{E_{vbiogas}}{V_{biogas}} = \frac{18 \times 1 \text{ ml}}{200 \text{ ml}} = 9\%$$

Since the net normalized volume of methane, $V_{methane}$, results from a formula in which the ambient pressure, the temperature, and the percentage of methane are all factors, each one measured with a given error, then the total error of $V_{methane}$ is as follows:

$$e_{Vm} = e_{barometer} + e_{thermometer} + e_{gas\ analyzer} + e_{biogas}$$

The error of the gas analyzer is not explicitly stated, but its brand and model are quoted in the paper. Checking the manufacturer's catalog, we find that the maximum error after a calibration is 2%. The researchers stated in the paper that they performed a calibration with 60% methane before taking each reading, so we can deduce that 2% error in the gas composition measure was a constant throughout the test duration. The error of the barometer is not specified, nor is the barometer brand and model quoted, but we can assume that a reputed research center has laboratory-grade instruments. The absolute error of an electronic barometer for laboratory and meteorological use is about 20 hPa, so we can estimate that the relative error was roughly 2% throughout the duration of the test. The temperature accuracy of the thermostatic cabinet is stated as ±0.5°C, hence the relative error when measuring 37°C is 1%. The total uncertainty of the net normalized methane volume is then

$$e_{Vm} = 2\% + 1\% + 2\% + 9\% = 14\%$$

Observe that the paper provides no information about the gas production of the blank, which contributes to amplify the error, as explained in Chapter 2. We will assume as simplification that the researchers employed a preincubated inoculum, so that no error amplification needs to be considered. Furthermore, the paper does not provide information about the moisture content of the measured gas. At the temperature of 37°C, almost 13% of the gas volume in the syringe is occupied by water vapor. We will assume for simplicity that the researchers considered this systematic error and corrected it accordingly, since the paper states that a blank was employed. Anyway, given such high instrumental uncertainty, we can question the results published from a formal point of view: the authors of the paper should have stated their results according to ISO GUIDE 98-3:2008. The same defines the maximum number of decimal figures when expressing the results of calculations derived from measured values, which that in this case is just two decimals:

$$V_{methane} = 0.176 \text{ Nm}^3/\text{kg VS} \pm 14\%$$

$$V_{methane} = 0.17 \pm 0.02 \text{ Nm}^3/\text{kg VS}$$

In other words, in this test it is impossible to discern if the Hohenheim instrument and ancillary equipment measured 0.15 or 0.19 Nm³/kg VS, since any value within said range could be possible. Hence, expressing the average as 0.176 Nm³/kg VS has no physical sense because the sum of the error margin of the measures contributing to calculate the said value is bigger than the millesimal order of magnitude. Such misconception is a constant in all the biogas literature: researchers seem to believe that expressing results with point and many decimals makes the said data more accurate.

6.3.7.1 Conclusions

1. This example shows that the data published in the paper are very precise (only 1.7% dispersion), showing high procedural skills and laboratory virtuosity of the researchers, but for practical effects the result is not accurate (14% uncertainty).

2. The BMP values of manure with straw bedding from the five different stalls are shown in Table 6.4. We can observe that the variation coefficient of the five averaged measures, 5.1%, is still smaller than the instrumental error. The Cartesian doubt is hence legitimate: Is such variability the result of the different fodder and metabolic state of the animals, as claimed by the researchers, or is just a consequence of the high uncertainty of their measurement method? A sign that this suspicion may be true is contained in the same data provided by the researchers. Looking at Table 6.4, we can observe that three samples gave the same result, 0.17 Nm³/kg VS (we already demonstrated that the figures beyond the second decimal position

TABLE 6.4

BMP Values of Horse Manure with Straw Published by the Hohenheim Research Group and Further Analysis of the Author (Last Line)

Sample	BMP [Nm³/kg VS]	Variation Coefficient [%]
A	0.176	1.73
B	0.173	0.68
C	0.163	5.97
D	0.191	3.45
E	0.175	2.98
Average	0.176	5.1

are irrelevant), while the outliers, 0.19 and 0.16 Nm³/kg VS, are both within the uncertainty margin of the test.

3. From a practical perspective, a biogas plant manager intending to feed his digester with horse manure from a stall with straw bedding may discard the outliers resulting from the paper and calculate his digester's ration with the representative average, 0.17 Nm³/kg VS, or adopt a conservative attitude and employ the lowest value, i.e., 0.15 Nm³/kg VS.

A final tip for those readers who eventually plan to employ horse manure containing straw bedding: Figure 6.12 shows a test performed by a biogas plant manager under the author's supervision. The instrument employed was an AMPTS Light with stirred 2-l batch reactors. Being a test meant for practical purposes and not for research, a simplified protocol was adopted, consisting of a single sample and a blank. The inoculum was preincubated for just a few days, hence the total uncertainty of the instrument is just 1%, but the amplification factor resulting from the inoculum's background activity was 4, so the overall uncertainty of this simple test is ±4%. It is not possible to analyze a standard deviation, because the test was carried without replicates, but the sample amount (120 g, no treatment at all) and the effective volume of digestion (1800 ml) are by far more representative of the real plant operation than the Hohenheim protocol. The resulting BMP was 0.174±0.007 Nm³/kg VS. Coincidence or just correct application of the norms on measure uncertainty? Note that, even being a test performed in an industrial environment by an unskilled operator, in this case the third decimal position is relevant, since it is the same order of magnitude of the error.

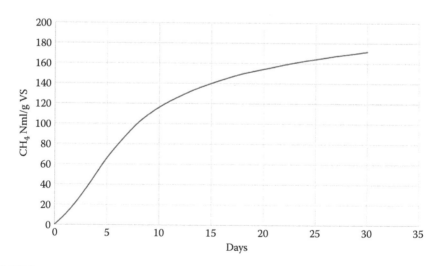

FIGURE 6.12
BMP_{30} of horse dung containing straw bedding performed by a biogas plant manager in his own laboratory, under the author's supervision.

6.4 Norms on the BMP Test Procedure for Industrial Biogas Plants

6.4.1 An Overview of the German VDI 4630/2014

We already presented in Table 3.2 a list of some norms on AD tests, pointing out that they are all focused on assessing the "ultimate anaerobic degradability" of wastewater, sludge, and urban waste. Therefore, their practical application to agricultural biogas plants is neither immediate nor pertinent, since the bacterial populations constituting the inoculum and the scope of the tests are different for both industrial sectors. In the waste treatment industry, the scope of AD is reducing as much as possible the pollutant OL and the volume of waste to handle, while the scope of all agricultural biogas plants is maximizing the energy production. Germany has the merit of being the first country in publishing a specific norm for the commercial biogas sector, with some focus on agricultural biogas plants running on energy crops. This norm, called VDI 4630, was first published in 2006 and reviewed in 2014. The most widely known version is the 2006 one because it is available in a bilingual German/English edition. The version commented here is the 2014, available only in German.

The scope of the norm is providing standard methods and criteria for the conduction of biological tests. Since commercial biogas plants began gaining economic importance throughout the world, the main problem that plant managers had to face was the large variability of the BMP values measured by different laboratories, plant constructors, and researchers. The said variability of the measure BMP values may sometimes lead to legal disputes, for instance, between plant owners and plant constructors or between plant owners and biomass vendors.

6.4.2 IWA's Proposal of Standardized BMP Test

The Anaerobic Digestion Experts Group of the International Water Association (IWA) published a draft of standard on September 19, 2016. Such proposal tends to overcome the worst problem all biogas industries and academic researchers face: the high dispersion of values resulting from BMP tests performed by different laboratories. The method described in the proposal will be validated during 2017 with a ring test that will involve a large number of biogas research laboratories.

6.4.3 The Draft of Italian Norm E0209F670 UNI/TS

At the moment of writing these lines (May 2017), the draft of norm is in its public survey phase. Depending on the type and quantity of remarks received by UNI until late June, the draft will be either approved or sent back to the commission for improvement. The Italian laws foresee that, after internal revision by the national standardization body (UNI), the draft of norm must be openly published for public survey. All the comments received after 60 days must then be analyzed and the draft corrected, if only minor corrections are necessary, or sent back to the expert's commission for reviewing critical aspects. Once the standard enters in force, a test period of 2 years allows the standardization body to collect any issues arising from the application of the standard. After said period, the standard is either reviewed or confirmed.

In broad lines:

1. The norm will apply only to batch tests, wet digestion (5% maximum solids), only to determine the BMP of solid waste and energy crops for energy purposes. Determining the residual greenhouse gas effect of biomass, dry digestion processes, and tests with continuous reactors is all out of the norm's scope.

2. The protocol defined in this draft foresees as minimum experimental setup, one blank reactor and two sample reactors. The minimum reactor volume is 500 ml, so tests with syringes and "serum bottles" are excluded. Tests can be performed either with barometric or volumetric methods. The (proposed) maximum acceptable error for barometric instruments is 2.5%, while no agreement has been reached yet for volumetric instruments.

3. Inoculum must be filtered with a mesh in the range of 1–5 mm in order to separate macroscopic solids and preincubated at least 5 days or until reaching a given specific daily production.

4. Trace elements must be added if the laboratory is not 100% sure of the inoculum's "health."

5. Control tests with cellulose or acetate are not mandatory with each experiment, but should be performed periodically as internal check of the laboratory's accuracy. They are recommended when the laboratory must test a biomass using an "unknown" inoculum, i.e., an inoculum whose actual methanogenic and hydrolytic activity has not been tested before.

6.4.4 Critical Analysis of the Flaws in the Existing Norms

All three cited norms and drafts of norm contain logical flaws and recommendations based more on "we have always done it in this way" than on sound metrology principles and open-minded experimentation.

6.4.4.1 *The VDI 4630*

The first puzzling aspect when reading norm VDI 4630 is its extreme complexity. Reading between the lines, it appears evident that its redaction was conditioned by the two different schools of thought that have always influenced the concept of BMP itself. The first school of thought is the one we could call "theoretical," since it derives the BMP from chemical analysis and stoichiometric calculations. The second school of thought is the "experimental" one, since it calculates the BMP from biological tests. Paragraph 4 of VDI 4630, *Characterization of the substrate*, was obviously influenced by the theoretical school of thought, since it defines a series of preliminary chemical analysis that should be performed, with the scope of assessing the digestibility of the biomass. Said analytical procedures are in general complex, so it is implicit that they must be carried out in a laboratory adequately equipped. Hence the possibility of self-management of the biogas plants is excluded, or at least strongly impaired, if the plant manager wants to fully comply with the norm. The limitations of such theoretical methods were already presented in Chapter 1, Section 1.4.3. In the author's personal opinion, including such useless tests as a norm prescription is just an attempt to discourage the owner to self-manage the biogas plant, in favor of the companies that offer biological management services, one of the pillars of the German biogas industry. It is spontaneous to ask oneself: What is the sense of performing a lot of analytical tests to check if a feedstock is degradable, and when it is easier to load a reactor with a sample and check if it produces gas or not?

In the same paragraph 4 of the norm, the reader can find an academic preconception that is not supported by an adequate experimental evidence: the statement "glycerol can be classified as fermentable with no limitations." Such statement is not universally valid. The tests performed by the author on crude glycerol (the by-product of the biodiesel industry), already presented in Figure 6.9, show that the said substrate has a strong inhibiting power. Nevertheless, it is often promoted commercially as "fodder for biogas plants" (at least in Germany and Italy). The reasons are many: Quite often, the caustic soda employed for the biodiesel production is neutralized with hydrochloric acid, forming common salt; second, crude glycerol can contain up to 12% of soap, and finally, the first step in glycerol's anaerobic degradation is propionic acid. All three substances are strong inhibitors of the anaerobic process.

The reader should be aware that the biodiesel industry is very strong in Germany, and the pressure to get rid of the glycerol is high, because it is a by-product with little demand. Again, reading between the lines, the classification of glycerol as "fermentable with no limitations" seems more as a concession to a specific industry, applicable only to Germany, than a rule for universal application, as should be expected from a norm. Not all anaerobic plants are capable of digesting glycerol "without limitations," so laboratory tests are necessary to check the inoculum's suitability in order to avoid inhibition and loss of profit.

In the author's modest opinion, including in the norm the juridical classification of biomass in waste, by-products, wastewater, animal excreta, and "renewable raw material" (a.k.a. "energy crops") is just a useless digression. From a technical point of view, the digestibility of a given substrate is independent from the classification imposed by the "eurobureaucrats"—by the way, a very questionable legal classification. Hence, the said paragraph can serve as a guideline in Germany and other European countries, only from a legal point of view, but has no scientific base and cannot be adopted elsewhere as a rule.

Another aspect, purely formal but not less important because of that, is the validity of the measure units adopted as standard for the calculations and the expression of the results. The VDI 4630 adopts the liter as measure unit of gas volumes. From the formal point of view, this is not correct because liter and its submultiples are not SI units (international metric system). There are norms on how to write norms, and the German experts who redacted the VDI 4630 seem to have ignored them. But the *Bureau International des Poids et Mesures* is very clear on the correct definitions of the measure units to express laboratory results, and the rules are public domain:

> Cubic decimetre and litre
> The Comité International des Poids et Mesures recommends that the results of accurate measurements of volume be expressed in units of the International System and not in litres. Reference: http://www.bipm. org/en/CIPM/db/1961/0/)

From a practical point of view, the production of biogas—or biomethane—is measured in Nm^3 (m^3 in normal conditions, i.e., 0°C and 101.3 kPa). The liter cannot be considered a measure unit derived from primary units (m, kg, and s) and is deemed obsolete. Hence, the correct unit to define volumes within the text of a standard is the cubic meter (and its symbol, established by the BIPM, is m^3). Submultiples of the m^3 are acceptable. BMP values should hence be expressed in Nm^3/kg VS (primary SI units) or Nm^3/ton VS (acceptable derived SI units).

Chapter 5 of the VDI 4630 standard is dedicated to the methods for sampling the different kinds of substrates for industrial biogas production and the preparation of the samples prior to laboratory tests. The importance of the sample's representativeness for obtaining reliable tests results is undeniable. Nevertheless, the contents of Chapter 5 seem somehow difficult to apply in practice. For instance, defining that sampling manure from a stall for a simple BMP test requires sending a specialized technician with a background in sampling methodologies—obtained in special training courses— appears as typical European overregulation. Furthermore, assuming that sampling manure is such a delicate and difficult task that deserves having a diploma, the "special courses" should be defined: Who can be the professors? What kind of school must provide such training? What is the duration and content of the courses?

The guidelines for the preparation of the sample prior to the test are reasonable: coarse materials should be grounded to 10 mm or less. The next methodological steps—separating and weighing the inert materials, like small stones or sticks, sieving with a 10-mm mesh, breaking big lumps...—are all scientifically correct, but put altogether they end up being too laborious and hence not practical. The following sentence appears especially grotesque: "The reduction of the average grain size must be performed with a sieve, pushing the sample with a hard wood stick having a diameter in the range 10–50 mm." It is impossible to figure out how employing other trituration means, or a stick made of a different material or with different diameter, could have an influence in the result of a BMP test.

Chapter 6 of the VDI 4630 is a collection of quotations of other DIN standards (24 in total) and furthermore to methods that are not normalized, though clearly related to products "Made in Germany." Among the latter, the VFA/TA (FOS/TAC) test is already discussed in Section 6.2.2.

Considering that standards are not free and that the average cost of a DIN or ISO publication is about 120 €, a laboratory willing to perform BMP tests fully compliant with VDI 4630 standard should spend at least 2.800 € in norms to study, plus all the corresponding tools and instruments for each of the 24 additional tests. Again, such overregulation is a penalty for small companies and tends to favor the already established (big) laboratories, both private and public.

Chapter 7 specifies how to perform biological tests and contains some contradictions. On one side, it states the use of airtight reactors, preferably made of glass, having volumes in the range of 0.5–2 l, built according to DIN 38414-8 or DIN EN ISO 11734 standards. Checking the said norms, it turns out that their scope is testing sewage water sludge, and the reactors specified do not have stirrers (they are either eudiometers or barometric reactors). While the lack of agitation is not a big problem when testing wastewater or sewage sludge, because the said substrates have a very low content of solids and thermal convection is usually enough to grant enough mixing, agricultural and municipal waste biogas plants work with high concentrations of solids, hence the stirring plays an important role in the digestion of biomass. Ignoring the said fact in the laboratory tests makes it difficult to apply their results to the real-scale biogas plant, hence the use of reactors having some sort of stirring should be mandatory. Figure 6.13 is quite eloquent: a test was performed on microcrystalline cellulose. One reactor was stirred (60% power, 2 min on/4 min off), while the other was not. The error of the test was 1% because the inoculum's background production was negligible. The stirred reactors produced 6% more methane than the unstirred ones. The kinetic constants of both curves are quite different too, although the lag phase is the same for both.

At this point a trivial, yet legitimate doubt arises: What is the use of the VDI 4630 if it redirects to other (existing) norms that are meant for testing sewage sludge? In the author's personal opinion, the said test protocols are not useful for the management of agricultural or municipal waste biogas plants, because they not reach the main scope of the VDI 4630 standard: "Defining a method that reduces the high variability of the results of BMP tests." The reason why the norm in question fails to reach its scope is that it tries to harmonize its content to other existing norms, losing specificity

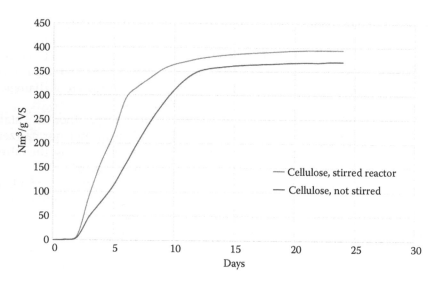

FIGURE 6.13
Influence of the stirring in the final BMP value.

by admitting any sort of reactor and measure instrument, regardless of its accuracy or suitability for the peculiar features of high-solids sludge. Such contradiction appears few lines after the definition of the reactors, where the norm makes an exception to itself and declares that the Hohenheim system, based on sets of 100-ml syringes, is acceptable. There is no scientific justification to such an exception, except for a self-referenced paper published by the Hohenheim University itself. We have already seen in Section 6.3.7 that the accuracy of measuring with such system is very low. We also commented that Chapter 5 of the VDI 4630 is devoted to the method of sampling and ensuring the representativeness of the sample. It is difficult to figure out how measuring with a syringe that cannot be loaded with more than 1 g of lyophilized and pulverized sample can ensure the representativeness of the latter. Hence, including such exception in a norm looks like a kind of protectionist concession to favor the "Made in Germany" brand.

Another critical aspect is the lack of compliance of Chapter 7 to the ISO/IEC Directive, Part 2—*Principles and rules for the structure and drafting of ISO and IEC documents.* This directive defines the method and protocol for redacting technical (ISO) norms and recommends that, in case two or more commercial methods are available for performing a given test or laboratory measure, the most accurate of them and its maximum admissible error margin must be defined. The error or uncertainty of measure is especially important in the case of the BMP assay, because the measured values have an economical relevance, as explained with the following practical example. Suppose a biogas plant manager needs to determine the maximum price that a given biomass lot is worth. The maximum affordable price is a function of the net methane production under anaerobic conditions which, just for common sense, should be as close as possible to the real plant's operational conditions. According to VDI 4630/2014, an eudiometer, a Hohenheim syringe carrousel, a barometric reactor with no stirrer, and a volumetric counter attached to a stirred reactor are all equally admissible instruments. Nevertheless, most of them measure in conditions that differ radically from those of a real-scale biogas plant, and each of them has a typical error margin, so a direct comparison of the test results is impossible. In other words, since the norm does not establish how to calculate the measure error range, each laboratory employing one of the test systems described earlier will obtain a single value as result, hence each laboratory can claim that its result is correct, even if it is wrong—or at least subject to a large margin of uncertainty. In any case, the VDI 4630 norm establishes that the results must be expressed in Nl/kg VS, but does not minimally consider the normalization error. According to a research carried out at the University of Lund (Södeborg et al., 2014), when measuring the BMP with instruments lacking the feature of normalizing in real time, as for instance, analogic barometers, eudiometers, and Hohenheim syringes, in the most extreme cases the normalization error can reach 80%.

Section 7.3.1 of the VDI 4630 norm provides very detailed formulas for calculating the BMP, but only valid for barometric instruments. These formulas include the correction of the moisture in the gas, but neglect the error in measuring the methane percentage and the error of the pressure gauge itself. We have already demonstrated in Chapter 2, Section 2.2.4.2, that the error induced by the gas moisture can be eliminated by simply ensuring that the headspace is the same both in the blank and in the sample reactors. Together with the lack of stirring, the error in the measure of the methane percentage is one of the weakest points of the barometric method—measuring the percentage of methane with high accuracy requires very expensive laboratory instruments, either gas chromatographer or infrared spectrometers, and time-consuming calibration before each test.

The same paragraph contains some correction formulas that must be applied when measuring the BMP of silage. From the scientific point of view, such approach is correct, but adds useless complexity to the test, increasing the cost of the biological management of the biogas plant.

The most disconcerting and metrologically wrong aspect of the VDI 4630/2014 is its Section 7.3.3: *Validity of the measures.* According to it, just performing triplicate tests and checking that the dispersion of the single values is <15% of the average (maximum 20% in the case of heterogeneous substrates, like municipal solid waste) is enough to ensure the reliability of the result. It is astonishing that in the country where the standardization of products and processes has its oldest tradition, a technical committee confuses two different concepts of metrology: precision (i.e., low dispersion of the measured values) and accuracy (i.e., the discrepancy between measured value and true value). Such confusion is even more difficult to understand if we consider that there are already specific international standards to define precision and accuracy of a test or measure (see footnote at the beginning of Section 6.3.7.).

Chapter 9 of the VDI 4630 is dedicated to continuous tests. In spite of the same being useful for some aspects of the biogas plant management, e.g., quick restart strategies or transition from a given diet to another, continuous tests require much labor over relatively long time spans, so they are more suitable for academic research purposes than for industrial applications. In the author's modest opinion, the preparation and conduction of continuous tests should be the subject of a specific norm.

Annex H of the standard shows an example of calculation, which at the same time is an example of academic presumptuousness and ignorance of the existing standards. In Table H1 of the VDI norm, the daily gas productions are expressed with dot and one decimal. We must remember that under normal test conditions and employing the best instruments in the market, it is difficult to reduce the test uncertainty under 3%. Hence, expressing a measure like, for instance, 90.1 Nml/g VS has no physical sense, since the true value could be any figure between 87 and 93, so the decimals are irrelevant. On the other side of the Atlantic, laboratory guidelines defined by the US NREL (National Renewable Energy Laboratory), although not having the status of national standard, are very clear: *A final result should never contain any more significant digits than the least precise data used to calculate it.* Such guidelines are among the first results in any search on the Internet and can be downloaded for free (http://www.nrel.gov/docs/gen/fy08/42626. pdf). Amazingly, the experts who wrote the VDI 4630/2014 seem not to have considered these guidelines as valid, or even worse, they have ignored their existence.

6.4.4.2 The IWA Guideline for the BMP Assay

The introduction of the document states that its scope is to overcome the high dispersion of BMP values encountered in the so-called interlaboratory or ring tests, a scope that VDI 4630 has failed to reach. In the author's modest opinion the IWA protocol will fail too. The reason was predicted by Henry Ford almost one century ago: "If you always do what you've always done, you'll always get what you've always got." The IWA draft protocol persists in the same misconception of the VDI 4630: confusing instrumental error of the measure with dispersion of the replicates. According to the IWA proposal, the acceptance criterion of a test is the standard deviation of the replicates. For instance, the test described in our example in Section 6.3.7 would be acceptable according to the said criterion, while we have demonstrated that it has at least 14% uncertainty.

Another misconception is defining as acceptable any inoculum that reaches the prescribed minimum BMP of reference substrates like acetate and cellulose. We have seen in the real-life example presented in Chapter 4, Section 4.3, that reaching a minimum BMP with some reference substrate is a necessary but not sufficient condition for assessing the inoculum's quality. The suitability of an inoculum for digesting any unknown biomass is not only having the capability, but being robust enough to carry out its task, as will be explained later with an example.

The consequence of the said flaws is that the ring test scheduled for 2017 will most probably yield the same result of all precedent ring tests on BMP: in the best case, the interlaboratory differences will be at least 15% because of the cumulated instrumental errors of some laboratories, plus eventually further 8%–20% because of overdosing trace elements or using "weak" or too starved inoculums, plus eventually further 3% in those laboratories employing reactors with no stirring.

Another commonplace contained in IWA's proposal is the hyped importance given to flushing the head volume of the reactors:

> During test preparation, flushing should be done with a mixture of N_2 and CO_2 that contains a similar share of CO_2 as expected in the produced biogas (e.g., 20%–40% CO_2; rest as N_2; v/v) to avoid a disturbance of the carbonate balance (Koch et al.). Flushing with pure N_2 should be carried out with care and only used for small head-space volume reactors.

Figure 6.14 shows a comparative test of cellulose digestion (with an AMPTS, hence "small headspace volume reactors"). Three reactors were flushed with propane (a hydrocarbon more similar to methane than nitrogen) and the others were left unflushed. The difference is 3%, very near to the uncertainty margin of this test, which was 1%, so flushing or not flushing, or flushing with gases not containing CO_2, induces very little difference in the result. For sure, the difference is much smaller than all the other measure errors

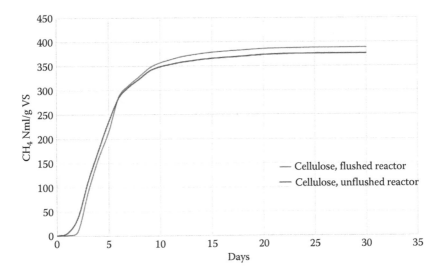

FIGURE 6.14
Influence of flushing the headspace volume.

affecting tests with unstirred and analogic reading apparatuses. At least with "small headspace volume reactors" and cellulose as reference substrate, flushing seems to be practically irrelevant. By the way, the unflushed reactors show a shorter lag phase, which demonstrates that a bit of oxygen favors the hydrolysis of cellulose at the expense of producing some additional mmol of CO_2. The AMPTS employed by the author to perform this test is sensitive enough to measure such small differences, but most other instruments in commerce would not be able to detect it.

The draft dedicates a paragraph to the inoculum-to-substrate ratio (ISR). It contains a generalization that cannot be considered an absolute rule, since it states,

> For easily degradable substrates where rapid accumulation of fermenta-
> tion intermediates such as VFAs could lead to inhibition of anaerobic
> digestion, an ISR greater than or equal to four should be applied. For less
> degradable substrates, such as lignocellulosic organic matter, an ISR less
> than or equal to one can be applied.

We have seen that glycerol and fatty substances usually require high I/S ratios in order to ensure that all the organic matter is adequately digested. Furthermore, VFA accumulation is usually an indication of either low acetogenic activity (in this case, propionic and butyric acids not being rapidly converted into acetic acid and cause inhibition) or low acetoclastic activity (in this case, the acetoclastic Archaea are outnumbered or partially inhibited and are not able to degrade the acetic acid at the same pace it is being produced). Hence, just increasing the I/S ratio instead of ensuring that the inoculum is suitable for digesting a given substrate is a practice that perhaps can be acceptable for academicians—since they only look to get the final BMP—but is useless and even dangerous for biogas plant managers because it does not allow them to diagnose eventual bottlenecks in the AD process. The suitability of the inoculum is connected to the next paragraph in the draft of standard: positive controls.

Positive controls are tests performed with reference substrates. The draft considers only cellulose and tributyrin or a mixture of both. Such specification is too reductive, because the fact that an inoculum is able to degrade cellulose or tributyrin does not mean that it will be able to degrade any other substrate. It is easier to understand this concept with an analogy. Imagine you need to select the athletes that will participate in the Olympic Games. Imagine that, in order to simplify the qualification procedure, the selection criterion is defined as just being able to run 100 m. First of all, if the maximum time for running 100 m is not specified, anybody in any discipline can compete, even persons not qualified as athletes. If the condition is set to running 100 m in 11 s, then a weight-lifting champion would probably be discarded, although being a champion is his specialty. Conversely, an athlete able to run 100 m in 11 s would qualify for competing as a swimmer or in weight lifting, even if not necessarily fit or trained for such disciplines.

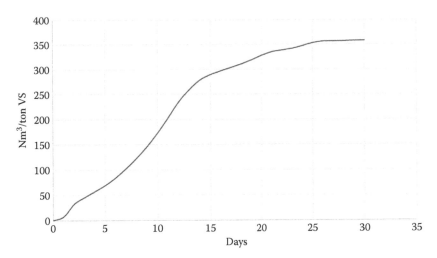

FIGURE 6.15
Anaerobic digestion of cellulose with a "weak" inoculum. Even if the final BMP reaches the minimum admissible value during the test with cellulose, such inoculum will probably yield underestimated BMP of other substrates.

Assessing the suitability of an inoculum for BMP tests is similar to selecting an athlete for a pentathlon competition: the candidate must prove not only being able to perform different activities, but also perform *all of them* in acceptably short times, but not necessarily record-breaking times.

Figure 6.15 is an example of an inoculum having some kind of biological problem that would anyway comply with the IWA guideline, since it reaches a BMP slightly higher than the minimum admissible (352 Ndm³/kg VS). It is evident from the shape of the cumulated production curve that such inoculum has problems because it reaches the plateau after 25 days while cellulose is usually degraded between 12 and 20 days. Furthermore, the curve presents several humps and a sudden plateau, while a good inoculum would degrade cellulose giving a nearly perfect sigmoid (compare Figure 6.15 with Figure 6.14).

Figures 6.16 and 6.17 show, respectively, the digestion of starch and gelatin with the same inoculum. The reader can draw his own conclusions.

Another common place about inoculum is its conservation: the draft of norm states that it must be kept at operation temperature, but if this is not possible, at least at room temperature, but <5 days. The experiment presented in Section 3.5 shows that such disposition may be a bit excessive since longer times at room temperature can be admissible.

6.4.4.3 The Italian Draft of Standard

The draft E0209F670 UNI/TS is the only one known to the author that includes a chapter on error propagation, although the same is defined as "informative" (i.e., full compliance is not mandatory). In September 2016, the

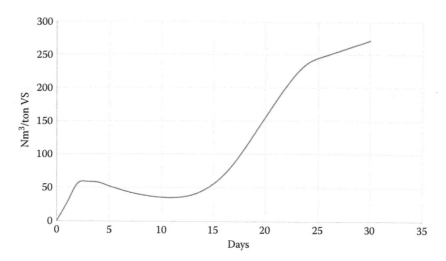

FIGURE 6.16
Digestion of starch with the same inoculum of Figure 6.15.

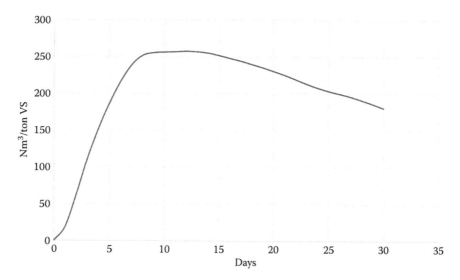

FIGURE 6.17
Digestion of proteins with the same inoculum of Figure 6.15.

Italian UNI adopted as national standard the translation of the ISO GUIDE 98-3:2008 (Guide to the expression of the measure uncertainty), with the name *UNI CEI 70098-3:2016* "Incertezza di misura—Parte 3: Guida all'espressione dell'incertezza di misura". So, hopefully the "informative" paragraph may become "normative" (i.e., mandatory).

Like in the IWA guideline, the kinetics of the AD is not considered as an indication of the inoculum's suitability, only its ability to reach the minimum defined value in the tests with cellulose and acetate.

The conservation of the inoculum when not in use should be done at 4°C.

The proposed procedure to flush the head space of the reactor is more complicated than the one proposed by the IWA:

> The test and blank reactors must be flushed with inert gas, preferably a mixture N_2/CO_2 (50/50 or 80/20 in volume). N_2 can be employed as alternative. The flushing must be maintained during at least 5 minutes at a flow equal to 1 headspace volume per minute, i.e. the total volume of flush gas must be equal to 5 times the headspace volume, flowing during at least 5 minutes.

We already discussed the scarce influence of the flushing, at least in reactors with small headspace. Furthermore, one could argue: If biogas is composed by methane and CO_2, why should the flushing be done with "inert gas?"

6.5 Conclusions

"There is no book so bad that it does not have something good in it." The maxim, coined by Miguel de Cervantes Saavedra in his famous work *Don Quixote*, is as valid for novels as for technical literature. The VDI 4630/2014 standard surely contains many positive aspects, as for instance, a full chapter on how to translate the results of batch tests to continuous industrial digesters. Nevertheless, in general lines, its practical application is too difficult and expensive for industrial biogas plants, its content is too verbose, rich in contradictions and conceptual metrology misconceptions. These latter make the results obtained with such protocol quite questionable and contradictory with the ISO Guidelines 98 on the measure uncertainty and expression of the results.

The Italian standard on BMP has the merit of being much more specific and easier to apply than the German one, but still has several points that could be improved, and is currently in the state of draft.

The IWA proposal of standard will be validated with an interlaboratory test during 2017. Since it contains the same logic flaws as VDI 4630 and "the literature" in general, it is very likely that the results of said ring test will continue to show differences in the range of 20%–30%, like already happened in all former ring tests. *Do what you've always done and you'll get what you've always got…*

At the time when the author is writing these lines (December 2016), there is still much work pending until reaching a satisfactory standard for performing the BMP test that finally eliminates disputes between the different stakeholders of the biogas industry and improves the quality of the scientific literature in general. The most difficult hurdle to overcome is the academic obstinacy, sometimes bordering arrogance, since each research group presents its own internal procedures and instruments as "the correct way," with the argument that such procedures and instruments lead to "results in line with those of the literature." During the times of Galileo, "the literature" said that the Earth was static at the center of the Universe. *Eppur si muove...* (Nevertheless, it moves), Galileo uttered.

From the author's practical experience, the only way to get reliable results from a BMP batch test is complying with the following simple rules of thumb:

1. Volumetric methods with real-time normalization and mechanical stirring must be preferred, since they mimic better the real operation of a biogas plant. Stirring must be gentle, but as continuous as possible.

2. Measure net methane directly (i.e., reactors connected to the measure instrument though a caustic soda filter).

3. Reactors must have at least 500 ml in order to accept a quantity of sample that can be deemed "representative." This is particularly important for agricultural biomasses, which are usually heterogeneous. Reactors bigger than 2 l bring no additional accuracy.

4. Use two blanks and triplicate sample reactors if these have less than 1 l capacity, and use one blank and two sample reactors if these have 2 l or more capacity.

5. Incubate the inoculum until its residual gas production is less than $0.03 \, \mathrm{Nm^3/kg}$ VS of inoculum·day in order to minimize the error amplification factor in the BMP formula.

6. The total cumulative error as well as the standard deviation of *both* blanks and samples must be checked, and outliers eliminated (better, repeat the test). A test is good (accurate *and* precise) if the total relative error and the variation coefficient are *both* less than 5%. A test is acceptable for industrial scopes if its relative error is smaller than 5%, even if the variation coefficient is bigger (average value accurate, but not precise). A test is doubtful if its relative error is bigger than 5%, even if the variation coefficient is smaller than 5%.

7. Check the experimental setup to find small gas leaks before starting the experiment.

8. If the inoculum's microbial activity is not known in advance, then a complete set of positive controls must be performed in parallel with the BMP measure of the sample biomass. Positive controls must be performed with acetate, cellulose, starch, protein, and fat, in order to check that all microbial groups are present and active. Positive controls can be performed in single reactors.

9. The quality of the inoculum is good if it passes a triple check:

   a. The positive controls must reach at least 80% of the BMP of each reference substrate.

   b. The shape of their cumulate volume curves must be clearly sigmoid.

   c. The acceptable time to reach at least 80% of the minimum admissible BMP must be the following: 2 or 3 days for the test with acetate, 4–7 days for the test with starch, 10–15 days for the test with cellulose, 10–20 days for the test with protein, 20–45 days for the test with vegetable oils or fats.

10. If the positive control curves are not sigmoid, and/or the acetate takes more than 3 days for complete degradation, then trace elements must be added to the inoculum, it must be degassed again, and the test repeated.

11. Flushing with N_2 (or any other inert gas except pure CO_2), CH_4, biogas from the plant, or LPG is advisable but not a critical issue, at least for reactors with small headspace. A volume of flush gas with at least twice the headspace volume should be enough.

12. pH is not such an important issue in agricultural biogas plants, since manure usually has more alkalinity than necessary.

13. Keep the same headspace and the same amount of inoculum in all reactors and blanks in order to eliminate the systematic error induced by the gas moisture or by the VS correction factor. If necessary, add distilled water to the blank.

14. Cumulated volume curves with two or even three plateaus (like the one discussed in Figure 4.3) are acceptable for mixtures of substrates. This is a normal feature of the batch test for compound substrates (sugars, protein, and fat), not a procedural error.

Bibliography

Abrol, I.P., Yadav, J.S.P., and Massoud, F.I., *Salt-Affected Soils and their Management*, FAO Soils Bulletin 39, Food and Agriculture Organization of the United Nations, Rome, 1988. http://www.fao.org/docrep/x5871e/x5871e00.htm#Contents.

Bakhov, Z.K., Korazbekova, K.U., and Lakhanova, K.M., The kinetics of methane production from co-digestion of cattle manure, *Pakistan Journal of Biological Sciences*, 17(8), 1023–1029, 2014.

BIPM (Bureau International des Poids et Mesures), Evaluation of measurement data—Guide to the expression of uncertainty in measurement, 1st edition, JCGM 100, Paris, France, 2008.

Bruni, E., Jensen, A.P., and Angelidaki, I., Comparative study of mechanical, hydrothermal, chemical and enzymatic treatments of digested biofibers to improve biogas production. *Bioresource Technology*, 101(22), 8713–8717. doi:10.1016/j. biortech.2010.06.108.

Burgstaler, J., Wiedow, D., Godlinski, F., and Kanswohl, N., Einsatz von Natriumhydrogencarbonat in landwirtschaftlichen Biogasanlagen, *Landbauforschung—VTI Agriculture and Forestry Research*, 4, 343–352, 2011. https://www. thuenen.de/media/publikationen/landbauforschung/Landbauforschung_ Vol61_4.pdf.

Choong, Y.Y., Norli, I., Abdullah, A.Z., and Yhaya, M.F., Impacts of trace element supplementation on the performance of anaerobic digestion process: A critical review, *Bioresource Technology*, 209, 369–379, 2016.

Dupont, Biogas enzymes brochure, http://www.dupont.com/content/dam/ dupont/products-and-services/industrial-biotechnology/documents/DuPont-BiogasEnzymes-brochure-web-EN.pdf.

FAO, Maize in human nutrition, FAO Food and Nutrition Series, No. 25, ISBN 92-5-103013-8, Rome, 1992. http://www.fao.org/docrep/t0395e/t0395e03.htm.

Garuti, M., Soldano, M., and Fabbri, C., Conducibilità elettrica, utile monitorarla nel digestato, article published by L'Informatore Agrario, 40, 2014.

Gustavsson, J., *Cobalt and Nickel, Bioavailability for Biogas Formation*, Linköping Studies in Arts and Science No. 549, Water and Environmental Studies, Linköping, Sweden, 2012.

Helffrich, D., and Oechsner, H., The Hohenheim biogas yield test: comparison of different laboratory techniques for the digestion of biomass, Landtechnik 3/2003.

Holliger, C., Alves, M., Andrade, D., Angelidaki, I., Astals, S., Baier, U., Bougrier, C., Buffière, P., Carballa, M., de Wilde, V., Ebertseder, F., Fernández, B., Ficara, E., Fotidis, I., Frigon, J.-C., de Laclos, H.F., Ghasimi, D.S.M., Hack, G., Hartel, M., Heerenklage, J., Horvath, I.S., Jenicek, P., Koch, K., Krautwald, J., Lizasoain, J., Liu, J., Mosberger, L., Nistor, M., Oechsner, H., Oliveira, J.V., Paterson, M., Pauss, A., Pommier, S., Porqueddu, I., Raposo, F., Ribeiro, T., Pfund, F.R., Strömberg, S., Torrijos, M., van Eekert, M., van Lier, J., Wedwitschka, H., and Wierinck, I., Towards a standardization of biomethane potential tests, *Water Science & Technology*, 2016. http://wst.iwaponline.com/content/ppiwawst/ early/2016/09/19/wst.2016.336.full.pdf.

ISO/IEC Guide 98-1:2009, Uncertainty of measurement—Part 1: Introduction to the expression of uncertainty in measurement. https://www.iso.org/ standard/46383.html.

ISO/IEC Guide 98-3:2008, Uncertainty of measurement—Part 3: Guide to the expression of uncertainty in measurement (GUM:1995). https://www.iso.org/ standard/50461.html.

Lebuhn, M., Liu, F., Heuwinkel, H., and Gronauer, A., Biogas production from monodigestion of maize silage–long-term process stability and requirements, *Water Science & Technology*, 58(8), 1645–1651, 2008.

Luna del Risco, M., Normak, A., and Orupõld, K., Biochemical methane potential of different organic wastes and energy crops from Estonia, *Agronomy Research*, 9(1–2), 331–342, 2011.

Michener, B., Scarlata, C., and Hames, B., Rounding and significant figures: Laboratory analytical procedure (LAP), issued July 17, 2005, Technical Report NREL/TP-510-42626, January 2008.

Mönch-Tegeder, M., Lemmer, A., Oechsner, H., and Jungbluth, T., Investigation of the methane potential of horse manure, *Agricultural Engineering International: CIGR Journal*, 15(2), 161–172, 2013.

Petta, L., Giuliano, A., Porzio, V., and Gamberini, P., Con il pretrattamento enzimatico cresce la sostenibilità della filiera, *Terra e Vita*, 29–30, 50, 2013.

Ploechl, M., Hilse, A., Heiermann, M., Quiñones, T.S., Budde, J., and Prochnow, A., Application of hydrolytic enzymes for improving biogas feedstock fluidity, *Agricultural Engineering International: CIGR Journal*, IX, 2009.

Södeborg, S., Nistor, M., and Liu, J., Towards eliminating systematic errors caused by the experimental conditions in biochemical methane potential (BMP) tests, *Waste Management*, 34, 1939–1948, 2014.

Verein Deutscher Ingenieure, VDI 4630-2014 Vergärung organischer Stoffe Substratcharakterisierung, Probenahme, Stoffdatenerhebung, Gärversuche (Entwurf—April 2014), ICS 13.030.30, 27.190. https://www.vdi.de/technik/fachthemen/energie-und-umwelt/fachbereiche/energiewandlung-und-anwendung/richtlinien/vdi-4630/.

7

Glossary of Terms and Abbreviations

AD Anaerobic digestion: A fermentative process in which microorganisms degrade organic matter, yielding methane and carbon dioxide as the ultimate products.

ADP Anaerobic digestion plant

Alkalinity: A measurement unit of a solution's resistance to pH dropping. It is proportional to the concentration of bicarbonates or carbonates. When expressed in equivalent mg of calcium carbonate, it is called Total Alkalinity.

AMPTS Automatic methane potential test system

Ash: The mineral fraction of organic matter.

Biological collapse: A condition in which one or more groups of microorganisms are inhibited, breaking the trophic chain of organic matter degradation. It features the accumulation of intermediate products (usually volatile fatty acids [VFA]) and limited or null methane production.

Blank(s): Reactor(s) containing only inoculum. The net production of methane is assumed to be the difference between the average production of the replicates minus the average production of the blanks.

BMP Biochemical methane potential: Net quantity of methane produced by anaerobic digestion. When followed by a suffix, the latter indicates the time at which the BMP was measured; e.g., BMP_{30} means the BMP resulting from a 30-days test.

BOD5 Biological oxygen demand at 5 days: It is a measure of the aerobic biodegradability of biomass. Sometimes, cited in the literature on wastewater treatment, but of little use for industrial biogas plants management, since there is no direct correlation between the aerobic and the anaerobic biodegradability or organic matter.

Buffer capacity: See alkalinity

COD Chemical oxygen demand: It is a measure of the quantity of carbon present in the organic matter. It can be employed instead of the volatile solids as a calculation parameter for the BMP test. See Chapter 8 for the equivalence between COD and VS.

CRT Cellular retention time: Similar to Solids Retention Time (SRT), but refers to the solids fraction containing live bacteria.

CSRT Continuously stirred reactor tank (aka CSTR = continuously stirred tank reactor): The most diffused type of digester in the agricultural biogas industry.

Digestate: The remaining sludge after the AD process, mainly consisting of indigestible matter (lignin, minerals) and live microbial biomass.

DM Dry matter: The fraction of matter excluding water. Synonym of total solids (TS).

d.w.: Dry weight

E Absolute error: By definition, the difference between the real value of a magnitude and the value effectively measured with a given instrument or device.

e Relative error: By definition, the quotient between the absolute error and the measured value.

Error propagation: When calculating a magnitude using two or more individually measured magnitudes as input data for a formula, the individual errors tend to add, resulting in a calculated value that usually has bigger error (uncertainty) than the individual measures from which it originates.

FOS/TAC: Commercial name of the VFA/TA ratio test, from the German acronym *Flüchtige Organische Säuren/Total Anorganisches Carbonat*.

Headspace: Portion of the digester, or laboratory reactor, occupied by biogas.

HRT Hydraulic retention time: The average time that a volume of liquid biomass or dilution water remains in the digester.

Inoculum: Biomass composed of live microorganisms that can degrade organic matter under given conditions. Inoculums can be aerobic or anaerobic, depending on the kind of process; e.g., yeast is the inoculum for fermenting either bread dough (aerobic) or beer (anaerobic). Ruminant's manure is both the inoculum and substrate for AD. Digestate is the usual inoculum for performing BMP assays.

LCFA Long chain fatty acids: Fatty acids have more than six carbon atoms. They are usually insoluble in water, tend to form emulsions, and tend to saponify. In general, LCFA are inhibitory for the AD process, and require a long time for their complete digestion (45–60 days).

Mix ratio: Intensity of stirring in a CSRT reactor.

Nl or Nm³ Liters or cubic meters in "normal conditions": Since the volume of gases varies with temperature and pressure, it is necessary to define at which conditions a gas volume has been measured. In the European biogas industry, the norms define gas volumes at 0°C and at the atmospheric pressure 101.3 kPa.

ODM Organic dry mass: Synonym of VS, usually employed in the European technical literature (literal translation from the German Organische trocken Masse).

OL: Organic load = kg of VS (or COD) loaded to the digester/day

OLR: Organic loading rate = kg of VS (or COD)/m³ of digester/day

ORP Oxidation reduction potential (aka redox): A measure of the electrons exchange capacity between chemical species. The unit of measure is mV. If the ORP value is positive, the reactions are oxidative (in general aerobic); if negative, they are reductive (in general anaerobic).

pH Hydrogen potential: Nondimensional scale that measures the acidity or basicity of a solution. If pH = 7, the solution is neutral (distilled and degassed water). If pH < 7, the solution is acid (vinegar, beer, Coca Cola, lemon juice, demineralized water, hydrogen peroxide). If pH > 7, the solution is said to be basic or alkaline (caustic soda solution, tap water, sea water, ammonia, bleach, soap for dish washing machines).

Replicates: Each of the reactors containing a given amount of sample biomass, whose average methane production is employed to calculate the BMP.

Short circuit (more correctly, hydraulic short circuit): Fluid dynamics condition causing an HRT much shorter than the nominal. It is equivalent to a virtual reduction of the digester's volume.

Sl or Sm³ Standard liters or cubic meters: The concept is the same as that of Nl or Nm³, but defined as "standard conditions" at 20°C and 101.3 kPa. Not much diffused in the biogas industry, it is a standard from the natural gas industry.

Sludge: The digester's content, usually composed of live organic matter (bacteria and archaea, see inoculum) and partially degraded substrate.

SMA Specific methanogenic activity: Maximum amount of methane produced from sodium acetate (or acetic acid) per day, per g of VS of inoculum.

Substrate: The organic matter whose BMP is to be measured; feedstock of the biogas plant.

TOC Total organic carbon: Idem COD, but measured with a different method. Theoretically, $TOC = COD \times 12/32$.

TS: Total solids contained in the substrate (both organic and inorganic)

Turnover time: A coefficient defined as the quotient between the volume of the digester/recirculation flow. Employed as a control parameter in CSRT digesters using pumps as stirring system, desirable value in the range of 20–30 min.

Unit gas flow: A coefficient defined as the quotient between the injected biogas flow/volume of the digester [in CSRT digesters stirred by blowing biogas from the bottom, desirable values in the range 0.24–0.3 $(m^3/h)/m^3$ of digester].

Unit power: A coefficient defined as the quotient between the power at the stirrer's shaft/volume of the digester (in CSRT digesters with mechanical stirrers, desirable values in the range 5–8 W/m^3 for perfect stirring).

V Useful volume of the digester: Portion of the digester effectively occupied by the active sludge.

VG Velocity gradient: By definition, $VG = [P_{stirrers}/(V_{digester} \cdot \mu)]^{1/2}$ (where μ is the dynamic or absolute viscosity of the sludge).

Theoretically, the best way to define the intensity of stirring, but practically impossible to measure μ, because sludge is a non-Newtonian fluid. Ideal values in the range VG = 50–80 s^{-1}.

VFA Volatile fatty acids: Short chain fatty acids (<6 C atoms). The most frequently found in normal digestion processes are acetic, propionic, and butyric acids.

VS Volatile solids: Fraction of the TS assumed as being completely digestible by the microorganisms. (VS = TS − ash)

w.w. Wet weight: VS can be expressed as a fraction of DM or as a fraction of w.w. Throughout this book, VS are expressed as a fraction of w.w., except when otherwise specified.

8

Useful Tables for Quick Reference

This chapter contains a selection of tables already presented in the former chapters, and some additional miscellaneous information, for the reader's quick reference.

TABLE 8.1

pH of Some Common Substances

Substance	pH
Hydrochloric acid, 1 M solution	0
Acid battery	1.5
Gastric juice	1.0–2.0
Lemon juice	2.4
Coca Cola	2.5
Vinegar	2.9
Antibacterial intimate soap	3.5
Orange juice	3.7
Beer	4.5
Acid rain	4.5–4.8
Coffee	5.0
Tea, healthy skin, intimate soap	5.5
Deionized water at 25°C	5.5–6.0
Oxygenated water	6.2
Milk	6.5–6.7
Distilled water at 25°C	7.0
Healthy human saliva	6.5–7.5
Blood	7.40–7.45
Water in a swimming pool	7.2–7.8
Sea water	7.7–8.3
Alkaline soap	9.0–10.0
Ammonia	11.5
Bleach	12.5
Lye	13.5
Sodium hydroxide, 1 M solution	14

TABLE 8.2

Content of C, N, and C/N Ratio of Some Common Biomasses

Substance	N (% d.w.)	C (% d.w.)	C/N
Corn straw	1.2	68	56.6
Cabbage	3.6	45	12.5
Hay	4	48	12
Alfalfa	2.5	40	16
Leguminous hay	1.6	40	25
Cow manure	1.6–1.8	30–40	17–25
Sheep manure	3.8	49–76	13–20
Stall manure (mixed with straw)	0.8	22	27.4
Horse manure	2.3	57.5	25
Solid swine manure	2.8–3.8	23–38	6.2–13.7
Swine slurry	0.4	4	10
Oat straw	0.5	40	80
Wheat straw	0.5	50	100
Barley straw	1	48	48
Layer hen's dung	3.7–6.3	31–35	5–9.6
Tomatoes	3.3	41	12.5
Kitchen waste	1.9	54	28.6
Corn stalks	1.4	44	31
Clover	3	39	13

TABLE 8.3

Equivalence Ratios between Volatile Solids (VS) and Chemical Oxygen Demand (COD) of a Few Pure Substances Usually Employed as Reference Substrates in Laboratory Tests

Substance	Chemical Formula	g VS/g COD
Vegetal and bacterial biomass	$\approx 50\%$ C	1.3333
Glucose	$C_6H_{12}O_6$	1.0666
Sucrose	$C_{12}H_{24}O_{12}$	1.0666
Fructose	$C_6H_{12}O_6$	1.0666
Starch	$(C_6H_{12}O_6)_n$	1.0666
Cellulose	$(C_6H_{10}O_5)_n$	0.84375
Proteins (average)	55% C + 7% H + 17% N + 21% O	≈ 0.42
Acetic acid	CH_3COOH	0.93
Sodium acetate	CH_3COONa	≈ 0.93
Propionic acid	$C_3H_6O_2$	0.6622
Sodium propionate	$C_3H_5O_2Na$	≈ 0.6622
Butyric acid	$C_4H_8O_2$	0.5525
Sodium butyrate	$C_4H_7O_2Na$	≈ 0.5525
Vegetal oils (as oleic acid)	$C_{18}H_{34}O_2$	0.346

TABLE 8.4

Quantity of Daily Excrete and Approximate VS of Some Animal Species' Dung

	Daily Excrete (% of Live Weight)		Solids in Fresh Dung		Live Weight
Species	Manure	Urine	TS (%)	VS (%)	(kg)
Bovine	5	4–5	16	13	135–800
Buffalo	5	4–5	14	12	340–420
Swine	2	3	16	12	30–75
Rabbits	3	2	20	18	2–3
Sheep/goat	3	1–1.5	30	20	30–100
Hens	4–5		25	17	1.5–2
Humans	1	2	20	15	50–80

TABLE 8.5

Approximate Biochemical Methane Potential (BMP) of Some Common Agricultural Substrates (from Different Bibliographic Sources)

Substrate	Average BMP (NL CH_4/kg VS)
Swine manure	450
Bovine manure	250
Horse manure	460
Sheep manure	200
Stall manure (with straw)	225
Chicken dung	450
Corn straw	410
Corn silage	350–450
Rice straw	220
Grass	280–350
Vegetal waste	350
Sewage water	420

TABLE 8.6

Reference Substrates for Biological Activity Tests

Reference Substrate	BMP Range (min.–theor.)	Specific Bacterial Group
Cellulose	330–410 Nml/g VS	Hydrolytic bacteria
Starch	330–410 Nml/g VS	Hydrolytic bacteria
Glucose	330–373 Nml/g VS	Acidogenic bacteria
Casein or gelatin	370–470 Nml/g VS	Proteolytic bacteria
Propionic acid	330–350 Nml/g COD	Acetogenic bacteria
Butyric acid	330–350 Nml/g COD	Acetogenic bacteria
Long chain fatty acids (LCFA; vegetal oils)	700–1000 Nml/g VS	Lipolytic bacteria
Acetic acid	330–350 Nml/g COD	Acetoclastic Archaea
Sodium acetate	330–373 Nml/g VS (350 Nml/g COD)	Acetoclastic Archaea

N.B: The measured BMP is usually smaller than the theoretical one, because microorganisms convert part of the C present in the substrate into part of their own living biomass.

TABLE 8.7

Typical Measurement Error Margins of Some Standard Instruments Diffused in the Market

Instrument	Max. Relative Error	Remarks
AMPTS II	±1% (+ ega if NaOH filters are not employed)	Normalization in real time, resolution 10 mL, can measure net methane or total biogas
AMPTS light	±1% (+ ega if NaOH filters are not employed)	Normalization in real time, resolution 10 mL, can measure net methane or total biogas
Gas endeavour	±1% (+ ega if NaOH filters are not employed)	Normalization in real time, resolution 2 mL, can measure net methane or total biogas
μFlow Cell	±1% + e_{ga}	Normalization in real time, resolution 10 or 2 mL, generic gas flow meter
MilliGascounter Cell	±(3% + eb + et + ega if NaOH filters are not employed)	6% = min. normalization error assuming constant T and P for the normalization formula
Eudiometer and similar water displacement instruments	±(e_v + e_b + e_t)	e_v = error of the graduated cylinder or burette, calculated as the quotient between the cylinder's error class and the reading
Oxytop Bottle	±(1% + 1 hPa/P + e_T + e_v)	e_v = error in measuring the head volume of the bottle e_T = error of the temperature of the gas in the reactor's head space P = measured absolute pressure
Hohenheim syringes	±(5%–9% + e_b + e_t)	See an example of calculation in Chapter 6

e_b = error of the room barometer employed for the normalization of each reading; e_t = error of the room thermometer, idem; e_{ga} = error of the gas analyzer, including the uncertainty of the calibration mixture (when measuring total biogas production).

TABLE 8.8

Conversion factor between Nm^3 to Sm^3

$1\ Nm^3 = 0.947\ Sm^3$

8.1 Specific Weight of Silage as a Function of Its TS

The density of dry silage is fairly constant, independent of the ensiling conditions. The density of dry cereal silage is in the range of 235–245 kg/m³, while that of the dry grass silage is in the range of 250–260 kg/m³. The density of silage increases with its moisture content, but not linearly, because moisture causes the fibers to swell and hence changes the volume. To calculate the density of wet silage, multiply the corresponding value of dry silage defined here by the corresponding coefficient shown in Table 8.9.

TABLE 8.9

Density Correction Coefficients as a Function of Silage's Moisture

Silage moisture (%)	40	45	50	55	60	65	70
Multiply by	1.6	1.8	2.0	2.2	2.5	2.8	3.3

Example: Calculate the density of corn silage having 45% DM.

Moisture is equal to (100% – DM%), hence in this case it is equal to 65%. If you have not measured the density of dry silage, then you can assume 240 kg/m³ as an average value (the density of dry cereals' silage). Multiply by 2.8, obtaining then the wet silag's density as 672 kg/m³.

8.2 Using Wine Vinegar to Carry Out the SMA Test

White wine vinegar contains about 6% acetic acid, so it is necessary to add in total 4 g of caustic soda every 100 g of vinegar so as to neutralize it. After having added and perfectly dissolved 3 g, check with the pH meter that the resulting pH is less than 7, adding gradually the remaining caustic soda, and checking the pH, until it reaches at least 6.8. The resulting theoretical COD will be 6400 mg/0.1 l (i.e., 64,000 mg/l) and in the same way, the resulting theoretical VS will be 6 g/0.1 l, i.e., 6%.

8.3 Monitoring the Electrical Conductivity of the Sludge

As it happens with monitoring oxidation reduction potential (ORP) and pH, this method is of little or no use, as demonstrated in Chapter 6, but thousands of biogas plant managers believe in it with an almost religious faith. For such reason, the author decided to include this table (Table 8.10).

TABLE 8.10

Correlation between Electrical Conductivity and Methanogenic Activity

Electrical Conductivity (mS/cm)	Methane Productivity (%)
<12	100
20	85
30	65
40	32
>50	0

8.4 Reference Values for Checking the AD Process with the VFA/TA (FOS/TAC) Test

As demonstrated in Chapter 6, Section 6.2.2, the said test provides coherent values only when the process is healthy but it is absolutely unreliable when something is going wrong, since it measures all VFA *as if* they were pure acetic acid. A false supposition cannot lead to a true conclusion, if not by pure case, as explained in Chapter 6. Again, since many people believe in such tests with an almost religious faith, here is a reference table. In the author's opinion, sometimes, it may prove useful checking the VFA (FOS) and the TA (TAC) values independently. An increase in TA may indicate, for instance, an accumulation of minerals or ammonia (this is particularly true in plants that recirculate a consistent amount of liquid digestate as dilution water).

TABLE 8.11

Reference Values of the FOS, TAC, and pH, According to Daniel and Baumgartner

	FOS [mg/l]	TAC [mg/l]	FOS/TAC	pH
Normal operation	2100	9000	0.23	7.8
Beginning of the acidification	3300	8300	0.40	7.6
Acidified reactor	4800	6800	0.71	7.0
Biological collapse	7300	3800	1.92	5.6

8.5 VFA Profile

TABLE 8.12

Normal, Maximum Acceptable, and Limit of Biological Collapse of the VFA Most Commonly Found in the Sludge, in mg/l (from Several Sources)

VFA	Normal	Max, Acceptable	Collapse
Acetic	100–200	2,000–2,400	4,000–6,000
Butyric	10–1,800	4,000–6,000	8,000–12,000
Propionic	5–15	900–1,000	>3,000
Total VFA	<2,000	2,000 (when the substrate is cellulose) 4,000 (when the substrate is sugar)	

8.6 Oligoelements (a.k.a. Trace Elements or Micronutrients)

When the digester is fed with only cereals silage, the phytic acid contained in them chelates metals, turning them unavailable for the Archaea. It is then necessary to supplement the inoculum with such trace elements. Values in the Table 8.13 have been taken from several sources. It is highly advisable to keep the concentration of each mineral as near as possible to the minimum value, since they boost the activity of the acetoclastic archaea but at the same time, the hydrogenotrophic archaea could be inhibited by their presence. The safest action is performing a batch test in which a solution containing all the required trace elements will be added in different quantities to several test reactors, finding out which performs best.

TABLE 8.13

Reference Concentrations of the Different Trace Elements Necessary to Maintain the AD Process' Stability

Element	Concentration [g/m³ to mg/l]
Sulfur (S)	0.3–13,000
Iron (Fe)	0.3–4,800
Nichel (Ni)	0.005–5
Cobalt (Co)	0.001–10
Molibden (Mo)	0.001–50

8.7 Guidelines for the Determination of VS

TABLE 8.14

Minimum Quantity of Biomass Sample, Necessary to Keep the Error of the VS Below 1%, as a Function of the Absolute Error of the Scale, E_{scale}, and the Moisture Content of the Substrate Under Test

Substrate Under Test	Approximate Moisture	Minimum Sample Quantity for $e_{(VS)} < 1\%$
Dry: straw, meals, oleaginous cake	Up to 40%	$400 \cdot E_{scale}$
Semidry: silage, grass, urban waste, fresh manure	40%–80%	$1000 \cdot E_{scale}$
Humid: digestate, sludge, slurry	90%–95%	$3900 \cdot E_{scale}$

Index

Page numbers followed by *f* indicate figures; those followed by *t* indicate tables.